Managing Food Protection

James W. Kinneer, MA, CDM

DIETARY MANAGERS
ASSOCIATION

KENDALL/HUNT PUBLISHING COMPANY
4050 Westmark Drive   Dubuque, Iowa 52002

Dietary Managers Association
406 Surrey Woods Drive
St. Charles, IL 60174

Acknowledgments

The author wishes to express his appreciation to the professional staff and volunteer leaders of the Dietary Managers Association for their support of this book and their continued leadership in advancing the skills and knowledge of foodservice professionals. The author also greatly appreciates the time, insights and expertise shared by reviewers, including:

Susan Davis Allen, MS, RD
Southwest Wisconsin Technical College
Fennimore, WI

Raymond D. Beaulieu
Food Safety Specialist
Alexandria, VA

Linda S. Eck, MBA, RD, FADA
Northampton Community College
Bethlehem, PA

Susan Grossbauer, RD
The Grossbauer Group
Chesterton, IN

CONTENTS

PAGE

CHAPTER 1	Foodborne Illness: Causes and Prevention	1
CHAPTER 2	Purchasing, Receiving, and Storage	24
CHAPTER 3	Preparation and Service of Safe Food	42
CHAPTER 4	The Hazard Analysis Critical Control Point System	59
CHAPTER 5	Management and Personnel	80
CHAPTER 6	Cleaning and Sanitizing	99
CHAPTER 7	Physical Facilities and Equipment	119
CHAPTER 8	Safety Management	140
CHAPTER 9	Regulations, Inspections, and Crisis Management	158
APPENDIX A	Common Food Protection Terms	176
APPENDIX B	Suggested Responses to Case Studies	184
APPENDIX C	Review Questions Answer Key	189
APPENDIX D	References	190
INDEX		193

Foodborne Illness: Causes and Prevention

Chapter Objectives

- Define the problem of foodborne illness

- Identify the three major hazards to food

- Describe the growth conditions for bacteria

- Categorize common foodborne pathogens

- Cite preventive practices for avoiding foodborne illness

Foodborne illness is a major public health issue. A foodborne illness is a disease that is carried or transmitted by food. A *foodborne illness outbreak* occurs when two or more persons become ill after ingesting the same food and a laboratory analysis confirms that food was the source of the illness. An exception is a single incidence of chemically induced foodborne illness, which is considered an outbreak.

The Food and Drug Administration (FDA) estimates that every year, 24 to 81 million people become ill and 9,000 deaths occur as a result of microorganisms in food. The annual cost of foodborne illness is estimated to be between $7.7 billion and $423 billion. According to the Centers for Disease Control and Prevention (CDC), more than 250 different types of diseases are caused by contaminated food or beverages.

Foodborne Illness Causes

Food can become unsafe in a number of ways. Some individuals are more susceptible to foodborne illness than others. The very young, the very old and persons with a weakened immune system are considered to be high-risk groups for foodborne illness. There are many hazards in the environment that affect the safety and quality of food served. Hazards to food are present in the air, in water, in other foods, on work surfaces and on the hands and bodies of foodservice workers. These hazards to food can be divided into three major categories: biological, chemical and physical.

Biological Hazards. *Biological hazards* include bacteria, viruses, parasites and fungi. The greatest threat to food safety is bacteria. Bacteria are plentiful in the environment, and while some have useful purposes, others causes serious foodborne illnesses. Examples of techniques to prevent environmental contamination from biological hazards include: keeping food covered, following good personal hygiene habits, and maintaining a clean and sanitary foodservice operation.

Chemical Hazards. *Chemical hazards* include pesticides that are sprayed on food, preservatives used to maintain food, toxic metals in cooking equipment and utensils, and chemical cleaning materials used in the foodservice operation. Examples of techniques to prevent environmental contamination from chemical hazards include: purchasing produce only from suppliers who use approved pesticides, washing fresh produce to remove pesticides, using only equipment and utensils approved for foodservice use, and storing cleaning supplies away from food storage and preparation areas.

Physical Hazards. *Physical hazards* are foreign materials that enter food accidentally. Examples are glass fragments, metal shavings, staples from produce crates, and other objects that fall into food. Examples of techniques to prevent environmental contamination of food from physical hazards include: maintaining equipment in good repair, keeping food covered to prevent fragments from falling into it, and washing fresh produce to remove dirt.

Biological Hazards

To protect food in foodservice operations, it is important to understand the growth and control of microorganisms. Food protection efforts often focus on the control or destruction of microorganisms in food — and in everything that comes in contact with food.

Figure 1.1 ■ Food Hazards

Hazard	Examples
Biological Hazards	Bacteria
	Parasites
	Viruses
	Fungi
Chemical Hazards	Pesticides
	Metals
	Toxins
	Cleaning Chemicals
Physical Hazards	Dirt
	Hair
	Broken glass
	Metal shavings

While bacteria are most commonly cited in foodborne illness outbreaks, viruses, parasites and fungi also present a threat to the safety of the food supply.

Bacteria. Bacteria that cause foodborne illness are called *pathogens*. Pathogens may already be in the food supply or may be introduced at any point from the food processor to distributor to foodservice operation. Improper operating procedures and food preparation practices cause pathogens to survive and grow. In turn, these events increase the risk of foodborne

Figure 1.2 ■ Foodborne Illness Causes

Pathogens are naturally present in the food supply

⇩

Mishandling occurs during processing and preparation

⇩

Pathogens grow and other biological, chemical, and physical contaminants enter food

⇩

Food is consumed and foodborne illness results

illness. Examples of common pathogens are described in Figure 1.6.

Potentially hazardous foods are foods in which bacteria multiply and grow most rapidly. Meat, poultry, eggs, fish and seafood are examples of foods that are considered to be potentially hazardous, in either a raw or cooked form. Foods of plant origin that are heat-treated, cut melons and garlic/oil mixtures, are other examples. These foods contain the nutrients and water that bacteria require for growth. It is important to remember that bacteria can be transmitted to any food, so all food must be handled carefully.

Anaerobic bacteria grow in an environment with little or no oxygen. These bacteria might contaminate canned foods or modified atmosphere packaged products, or they may grow in the center of large quantities of food. Aerobic bacteria require air to grow. Many bacteria associated with foodborne illness can grow either with or without oxygen and are called *facultative*.

Foodborne illnesses caused by biological hazards may be classified as foodborne infections or foodborne intoxications. A *foodborne infection* results from ingesting food contaminated with bacteria that are in a vegetative or growing state. These active bacteria enter the body and continue to grow, resulting in foodborne illness. Bacteria in the vegetative state can usually be destroyed with heat in the cooking process.

Some bacteria are also capable of existing in a spore or dormant state. Spores are resistant to heat and can later return to a vegetative or growing state. Some bacteria produce a toxin which may not be destroyed by heat. Ingesting a food with such a toxin results in a *foodborne intoxication*.

Bacteria grow most rapidly in an environment that provides them with a nutrient or food source, moisture, warmth, neutral or slightly acid pH, oxygen at a level favored by the bacteria, and time. By controlling the time food spends in a temperature

Figure 1.3 ■ Conditions for Bacterial Growth: "POTTWA"

POTENTIALLY HAZARDOUS FOOD

Food must contain appropriate nutrients for growth. Bacteria generally prefer foods that are high in protein, such as meat, poultry, eggs, and dairy products.

OXYGEN

Some bacteria require oxygen (*aerobic*). Some bacteria cannot tolerate oxygen (*anaerobic*). Others grow with or without oxygen (*facultative*).

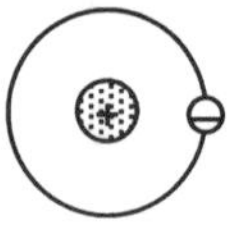

TEMPERATURE

Temperature is probably the most critical factor in the growth of bacteria. The temperature hazard zone of 41°F - 140°F (5°C - 60°C) is the range in which bacteria grow most rapidly.

TIME

A single bacterial cell can multiply into 1 million cells in 5 hours under ideal conditions. A general rule is that food in the hazard zone for 4 or more hours may be unsafe.

WATER

Moisture is measured based on water activity (available water or A_w). Pathogens grow in foods with a A_w greater than .85.

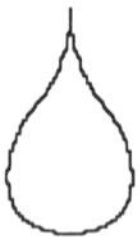

ACIDITY

Pathogens tend to prefer conditions that are near a pH of 7.0, but are capable of growing in a pH range of 4.6 to > 9.0.

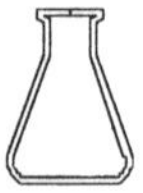

range favorable to bacterial growth, the foodservice manager can minimize growth during preparation and service. The temperature range at which most bacteria grow rapidly is called the *hazard zone*. The hazard zone is the range from 41°F - 140°F (5°C - 60°C). This temperature range is also commonly known as the *danger zone*. The terms *hazard zone* and *danger zone* can be used interchangeably. A potentially hazardous food that is allowed to remain in the temperature hazard zone for a cumulative time of four hours or more during all phases of receiving, storage, preparation and service may be considered to be unsafe.

Viruses. A *virus* is the smallest and simplest form of life known. Viruses are transmitted from one person (a foodservice worker) to another person (a customer) through food. Viruses do not grow or multiply in food. Foodborne viral illness is commonly caused by infected foodservice employees who fail to wash their hands. A foodservice worker infected with a virus is called a **carrier**. Infected workers should not be permitted to work in a foodservice operation. Additionally,

---

Figure 1.4 ■ The Safe Temperature Zones for Hot and Cold Food

The safe temperature zone for hot foods is 140° F (60° C) or higher.

Keep hot foods hot.

The hazard zone is the temperature range between 41° F (5° C) and 140° F (60° C), where bacteria grow most rapidly.

DANGER

The safe temperature zone for cold food is 41° F (5° C) or lower.

Keep cold foods cold.

Figure 1.5 ■ Guidelines for Handling Foods Containing Mold

Some foods that develop mold must be discarded. Other foods can be used, but first — cut out the mold and at least an inch of food around and under it.

Discard	Cut and Use
cucumbers	bell peppers
tomatoes	broccoli, cauliflower
spinach, lettuce, leafy greens	cabbage
bananas, peaches, melons	garlic, onions
berries	potatoes
breads, cakes, rolls, flour	turnips
cheeses like Brie or mozzarella	zucchini, winter squash
lunch meat and cheese (slices)	apples, pears
yogurt, tub spreads (cream cheese, etc.)	cheeses like cheddar or Swiss
(chunks)	
canned foods	
peanut butter	
juices	
most cooked leftovers	

Source: Center for Science in the Public Interest

all foodservice workers should observe proper hand washing and personal hygiene techniques to further prevention the possibility of transmission. Other sources of viruses are sewage-polluted waters and the marine foods harvested from them.

Parasites. A *parasite* is a small or microscopic organism that lives within another organism, called a **host**. The parasite can be transmitted from ani-mals to humans if food is not cooked thoroughly. Parasitic infection is far less common than bacterial or viral foodborne illness.

Fungi. Molds and yeasts are types of *fungi*. Some play a role in food processing. For example, yeasts are used in the production of bread, and molds are used in the processing of cheese. However, molds and yeasts can also be a threat to food safety. Molds grow

Figure 1.6 ■ Common Foodborne Pathogens

Bacillus cereus

TYPE	SOURCE	SYMPTOMS & COMPLICATIONS	PREVENTION
Bacteria	Milk, cereals, rice and dried food-stuffs	Nausea, diarrhea and vomiting are common symptoms. Onset is usually 1-16 hrs.	Keep food out of the hazard zone. Keep dry products away from moisture.

Campylobacter jejuni

TYPE	SOURCE	SYMPTOMS & COMPLICATIONS	PREVENTION
Bacteria	Raw meat, poultry and shellfish; contaminated water and unpasteurized milk	Muscle pain, headache and fever followed by diarrhea, abdominal pain and nausea. Children and young adults are affected more frequently.	Thoroughly cook meat and poultry. Use pasteurized milk and products. Control time/temperature. Use clean utensils. Use chlorinated drinking water.

Clostridium botulinum

TYPE	SOURCE	SYMPTOMS & COMPLICATIONS	PREVENTION
Bacteria	Grows in an-aerobic environment — canned and reduced-oxygen packaged foods, or large quantities of soups and sauces that have been improperly cooled	Double vision and difficulties in speaking, swallowing and breathing. May result in nerve damage and life-threatening illness.	Dispose of canned foods that are leaking or bulging, or are dented or damaged. Inspect vacuum-packaged foods for tears and temperature abuse.

Figure 1.6 ■ Common Foodborne Pathogens...*continued*

Clostridium perfringens

TYPE	SOURCE	SYMPTOMS & COMPLICATIONS	PREVENTION
Bacteria	Naturally present in the soil. Any raw food may contain the spore or bacteria. Called the "cafeteria germ" because many outbreaks result from from large quantities of food being held at room temperature or refrigerated too slowly.	Abdominal pain and diarrhea	Thoroughly cook foods containing meat or poultry. Maintain food at safe temperatures. Cool food quickly by dividing into small, shallow containers. Use ice as an ingredient, use an ice bath, or use a blast chiller. Reheat refrigerated foods adequately.

Escherichia coli 0157:H7

TYPE	SOURCE	SYMPTOMS & COMPLICATIONS	PREVENTION
Bacteria	Ground beef has been implicated in several severe outbreaks. Beef was often contaminated during processing and then mishandled during preparation.	Produces a toxin which causes hemorrhagic colitis. Symptoms include abdominal cramps, watery diarrhea, nausea, vomiting and low-grade fever. Acute kidney failure is a possible complication.	Cook ground beef to 155° F (68° C) for 15 seconds or more. Reheat properly. Use good sanitation practices. Refrigerate below 41° F (5° C).

Figure 1.6 ■ Common Foodborne Pathogens...*continued*

Listeria monocytogenes

TYPE	SOURCE	SYMPTOMS & COMPLICATIONS	PREVENTION
Bacteria	Commonly found in the intestines of animals and humans, in milk, soil and leafy vegetables. Also found in unpasteurized dairy products and processed meats.	In adults — sudden onset of fever, chills, headache, backache, abdominal pain and diarrhea. In newborns — may cause respiratory distress, refusal to drink, and vomiting.	Avoid unpasteurized milk and dairy products. Cook ground meats thoroughly. Maintain food temperature out of the hazard zone. Bacteria may grow at refrigerated temperatures, so follow "use-by" and "sell-by" dates on processed foods.

Salmonella enteritidis

TYPE	SOURCE	SYMPTOMS & COMPLICATIONS	PREVENTION
Bacteria	Often found in poultry and poultry products. Can survive in frozen foods and grow with or without oxygen. Common cause is poor personal hygiene or cross contamination.	Nausea, vomiting cramps and fever. Onset is within 12-36 hrs. Can be fatal to high-risk groups, such as the elderly.	Avoid serving raw eggs or undercooked poultry. Never pool eggs. Use only eggs with clean, intact shells. Sanitize work surfaces and utensils. Cook food thoroughly.

Figure 1.6 ■ Common Foodborne Pathogens...*continued*

Shigella

TYPE	SOURCE	SYMPTOMS & COMPLICATIONS	PREVENTION
Bacteria	An infected food-service worker (carrier) or contaminated water supply are frequently cited sources. Prepared salads, gravies, and milk have been implicated foods.	Onset is usually within 6-72 hrs. Symptoms include diarrhea, fever, chills, and dehydration.	Ensure a safe source of water. Restrict food-service workers with infectious disease from the foodservice operation.

Staphylococcus aureus

TYPE	SOURCE	SYMPTOMS & COMPLICATIONS	PREVENTION
Bacteria	Illness is caused by a toxin. Improperly sanitized equipment and poor hand washing practices culprits. Ham, cold meats, custards, and cream-filled desserts are implicated foods.	Nausea, vomiting, and abdominal cramps. Can be fatal to elderly, infants, and other high-risk populations.	Keep food out of the hazard zone by storing and cooking properly.

Figure 1.6 ■ Common Foodborne Pathogens...*continued*

Yersinia enterocolitica

Type	Source	Symptoms & Complications	Prevention
Bacteria	Can be found in meats, oysters, fish and raw milk. Poor sanitation and food preparation practices contribute to outbreaks.	Fever, headache, nausea, abdominal pain and diarrhea. Major complication can be unnecessary appendectomy.	Heat food thoroughly to destroy the bacteria. Practice effective hand washing.

Anisakis

Type	Source	Symptoms & Complications	Prevention
Parasite	Roundworm found in fish, transmitted through ingestion of raw or partially cooked marine foods.	Can be painful and is often misdiagnosed as appendicitis, Crohn's disease or gastric ulcer. Can require surgery to remove the parasite.	Cook marine foods thoroughly to destroy the parasite.

Trichinella spiralis

Type	Source	Symptoms & Complications	Prevention
Parasite	Roundworm found in pork. Illness caused by consuming pork that was not thoroughly cooked.	Muscle pain and fever.	Thoroughly cook pork to destroy parasites.

Figure 1.6 ■ Common Foodborne Pathogens...*continued*

Hepatitis A

TYPE	SOURCE	SYMPTOMS & COMPLICATIONS	PREVENTION
Virus	Virus is transmitted from one person to another through food. Water, shellfish and salads are frequently implicated foods.	Fever, abdominal pain, loss of appetite and jaundice. Typical incubation time is 10-15 days, but could be up to 50 days.	Practice proper hand washing and good personal hygiene. Restrict infected persons from foodservice operation.

Norwalk

TYPE	SOURCE	SYMPTOMS & COMPLICATIONS	PREVENTION
Virus	Contaminated shellfish or foods washed in contaminated water. Employees infected with the virus.	Nausea, vomiting, diarrhea and abdominal pain.	Avoid serving raw shellfish. Observe good personal hygiene and cooking practices.

well on most types of food and appear as brightly colored, fuzzy growth. Certain molds produce a toxin, which can result in a foodborne illness. Toxins produced by molds are called *mycotoxins*. Mycotoxins may be found in dry and/or acidic foods.

Yeasts are not a significant cause of foodborne illness, but can contaminate food and cause it to spoil. Yeasts prefer sweet, liquid foods and are well known for spoiling cider and fruit juices by fermenting them. An indication of yeast spoilage is a fermenting, alcoholic smell, and discoloration.

Chemical Hazards

Foodborne illness outbreaks caused by chemical hazards are less common than ones caused by biological hazards. Possible chemical hazards include pesticides, metals from cooking equipment, chemicals introduced from cleaning supplies used in the foodservice operation, and naturally occurring toxins.

Pesticides. Chemicals used to keep produce grown in the United States free from plant diseases and insect infestations are regulated by the government and should not cause foodborne illness when applied at recommended levels. But some foods originate from international sources, with less government regulation and monitoring. Wash all fresh fruits and vegetables to remove chemical residues before serving.

Metals. Metals that have been implicated in foodborne illness outbreaks include cadmium, antimony, lead, zinc and copper. Antimony, zinc, and copper may cause contamination after prolonged contact with acid foods. Carefully select foodservice equipment to control this hazard.

Cleaning Supplies. Products used to maintain a clean and sanitary foodservice operation may also cause contamination when improperly used or stored. Cleaning supplies and other toxic and poisonous materials must be stored separately from the food preparation area. Never use a food container to store chemicals, and never use chemical containers to store food. Containers used for cleaning — such as buckets and pails — should be clearly marked and dedicated for use in cleaning only.

Toxins. Toxins can occur naturally in food — such as fish, shellfish, mushrooms, and certain plants. A foodborne illness called ciguatera is caused by consuming fish that ingested toxins in algae. Scrombroid poisoning is the result of eating certain varieties of fish that release a toxin when they begin to spoil. Mushroom poisoning can result from eating certain varieties of mushrooms. To avoid foodborne illness from toxic species of mushrooms, purchase mushrooms only from an approved source.

Physical Hazards

Physical hazards, such as glass or metal fragments, can cause serious physical harm or injury. Physical hazards often enter food during processing or during the final stages of preparation. Foodservice workers must be trained in safe work practices to prevent physical contaminants from entering food.

Foodborne Illness Prevention

It is very important for foodservice operations to take steps to protect food, and to prevent and control hazards with proper food preparation and serving techniques. Examples of methods to prevent foodborne illness include:

■ **Purchase safe food.** Foodservice managers must carefully select an approved source for food, one that provides safe food. Schedule deliveries at a time convenient for the foodservice operation, and train all receiving staff in food safety principles. Inspect all deliveries to ensure that food meets quality and food safety expectations.

■ **Store foods promptly and properly.** After receiving, food should be stored quickly at a proper temperature in a clean storage area. Store all foods in tightly sealed containers. Dispose of food that shows evidence of damage from pests. Consult a pest control operator if you have evidence of a pest infestation.

■ **Prevent cross contamination.** Even safely cooked food can become contaminated if it comes into contact with raw foods. This type of contamination is called *cross contamination*. To prevent it, store all raw meat, poultry, seafood and eggs away from cooked foods and foods that will be eaten raw. Raw foods should never be stored above cooked or ready-to-eat foods because they may drip and cause contamination. Always wash fresh fruits and vegetables before cutting. Do not place cooked foods on cutting boards or work surfaces that were used to prepare raw, uncooked foods unless the surface has been cleaned and sanitized. Use clean and sanitized cooking utensils and service ware to prepare and serve all foods.

■ **Thaw food properly.** Food should never be thawed at room temperature. While at room temperature, food may enter the food temperature hazard zone, and rapid bacterial growth may result. Thawing under refrigeration is a desirable method that requires planning. Other acceptable methods include: under cold, running water; in a microwave as part of an uninterrupted cooking process; or thawing during the conventional cooking process.

■ **Cook food quickly and thoroughly**. The growth rate of foodborne patho gens can best be controlled by implementing time and temperature controls. Many raw foods, such as meat,

poultry, eggs, and unpasteurized milk, contain disease-causing bacteria and should not be eaten uncooked. To make sure these foods are safe to eat, cook them thoroughly. All staff members should be involved in monitoring food temperatures, and they should be trained in the proper procedures for recording food temperatures.

■ **Hold and serve food at a safe temperature.** When holding, maintain food in a safe temperature zone. Unserved food should be properly cooled and stored. When moving food from storage to preparation areas, label it to indicate the latest time for use.

■ **Cool cooked foods immediately**. Time is also an important control technique for food protection. As cooked foods cool, bacteria begin to grow. The longer food stands, the more bacteria grow. If storing hot foods, cool quickly by using shallow containers and reducing the volume of large food masses.

■ **Reheat cooked foods quickly and thoroughly**. Proper storage of leftover foods slows down the growth of disease-causing bacteria, but does not destroy them. To destroy any remaining bacteria, reheat leftovers to 165° F (74° C) for at least 15 seconds.

■ **Practice good personal hygiene.** Foodservice workers must wash their hands with soap and warm water for at least 20 seconds every time they prepare food and after every interruption — especially after using restroom, coughing, sneezing, eating, using tobacco, or handling a raw food. Restrict foodservice workers with infectious diseases from the foodservice operation.

■ **Use a Safe Source of Water.** The supply of a safe source of water is essential to food protection efforts. In addition to being served as a beverage, water is an ingredient in most foods and is used in the cleaning and sanitizing of equipment and utensils. Water must be *potable* (safe to drink) and supplied to the facility in a closed system approved by health regulations.

Key Concepts

■ Foodborne illness is a major public health problem affecting as many as 80 million people each year.

■ The hazards associated with food can be divided into three major categories: biological, chemical and physical.

■ Biological hazards include bacteria, viruses, parasites and fungi.

■ Bacteria are responsible for most outbreaks of foodborne illness.

■ Potentially hazardous foods are foods in which bacteria grow rapidly.

■ The acronym "POTTWA" describes growth conditions for bacteria: Potentially Hazardous Food, Oxygen, Time, Temperature, Water, and Acidity.

■ The temperature range at which bacteria grow most rapidly is called the hazard zone: 41° F (5° C) - 140° F (60° C).

■ Proper preparation and serving of food can prevent foodborne illness and its serious consequences.

Case Study

The health department is investigating an outbreak of food[…]
at a local day care center. The children were served a menu that
included tacos, tossed salad, fresh fruit, and milk. Terry, the cook, had
forgotten to pull ground beef to thaw under refrigeration a few days
earlier, so she allowed it to thaw on the counter overnight. After dividing
the partially thawed raw ground beef with her bare hands, she continued
preparations for lunch and chopped the ingredients for the tossed salad.
Using the same knife that she used to open the packages of ground beef,
she cut the fruit sections for the fruit salad.

☞ 1. What are the possible causes of foodborne illness outbreak
    at the day care center?

☞ 2. How could the outbreak have been prevented?

☞ 3. What are the possible consequences of the food safety errors
    made by the cook?

Review Questions

1. The biological hazard responsible for most cases of foodborne illness is:

    A. bacteria

    B. parasites

    C. viruses

    D. fungi

2. A small or microscopic organism that lives within another organism is called a:

    A. yeast

    B. mold

    C. fungi

    D. parasite

3. Anaerobic bacteria will NOT grow in the presence of:

    A. moisture

    B. acid pH

    C. oxygen

    D. warmth

4. The hazard zone for food temperatures is the range:

    A. 45° F - 145° F (7° C - 63° C)

    B. 41° F - 145° F (5° C - 63° C)

    C. 41° F - 140° F (5° C - 60° C)

    D. 45° F - 150° F (7° C - 60° C)

5. Which of the following is NOT a requirement for bacterial growth?

    A. warmth

    B. moisture

    C. oxygen

    D. alkalinity

6. Food may not be in the temperature hazard zone for more than:

    A. 2 hours

    B. 3 hours

    C. 4 hours

    D. 5 hours

7. Disease-causing microorganisms are called:

    A. spores

    B. pathogens

    C. toxigens

    D. facultative

8. Which of the following bacteria produces a toxin that causes hemorrhagic colitis?

    A. Listeria monocytogenes

    B. Staphylococcus aureus

    C. Escherichia coli 0157:H7

    D. Bacillus cereus

9. Which the following is a parasite?

    A. Hepatitis A

    B. Trichinella spiralis

    C. Salmonella

    D. Escherichia Coli

10. Which of the following is NOT a method of preventing foodborne illness?

    A. reheat quickly and thoroughly

    B. serve or store immediately

    C. cool quickly

    D. thaw gradually at room temperature

Purchasing, Receiving and Storage

Chapter Objectives

■ Distinguish between inspection and grading

■ Identify factors in selecting a safe food vendor

■ Describe procedures for verifying the safety of food during receiving

■ Identify safe temperatures for food storage

■ Describe the first in, first out (FIFO) inventory management method

■ Identify food protection factors throughout the steps of purchasing, receiving and storage

---

Ensuring food safety depends on selecting a safe food vendor and establishing sound purchasing practices. In addition, receiving and storage are critical. This chapter addresses safe practices for all these activities.

Purchasing

Food must be purchased from *approved sources*. An approved source is one that is inspected based on Federal, state, or local laws. The foodservice manager should include quality standards, such as grading, and wholesomeness indicators, such as inspection, in purchasing specifications. In addition, delivery times and intervals can also be included in specifications. The foodservice manager must exercise careful planning when purchasing food. Purchasing excessive quantities of food can result in spoilage, increased costs and an increased likelihood of foodborne illness. Perishable products — such as fresh produce, fresh seafood, fresh meats, eggs, and dairy products — should be purchased in a quantity that can be used within a very short time period. Non-perishable foods such as canned and frozen foods can be purchased in quantities that allow for longer storage times if desirable. Suggested storage times for many foods appear in Figure 2.7.

Inspection. All food shipped in interstate commerce (from one state to another) must meet the requirements of one or more Federal laws. The U.S. Department of Agriculture (USDA) has established uniform standards for state and Federally *inspected* meats, poultry, and eggs. Some products that are not shipped across state lines may have to be inspected by state programs, with their own standards — some higher than those of Federal programs. Foodservice operations may only purchase meat, poultry and eggs that have been inspected by the USDA or by a state department of health or agriculture.

Seafood and fish must be purchased from approved suppliers. The Public Health Service maintains lists of Certified Shellfish Shippers. Suppliers should be selected from this list or from state-approved lists. Identification tags for shellfish must be kept on file for 90 days after receiving shellfish such as oysters, mussels, and clams. In the case of illness, the identification tags can be used to trace a product back to the source and assist in determining if harvesting waters are polluted. Inspected food is considered to be safe for consumption, but inspection does not imply any other quality standard.

Grading. *Grades* are classifications of foods by a descriptive term or a number, to ensure the uniform quality and to give an indication of the desirable use. Most grades are assigned by government agencies which follow strict guidelines. Grades refer to attributes such as visual appearance, color, size, marbling, and uniformity.

Beef, veal, pork and lamb are graded by the USDA. For beef, the USDA grades are: prime, choice, select, standard, commercial, utility, cutter, and canner. The last three grades are not used in foodservice operations. A major distinction between inspection and grading is that inspection programs are mandatory for meat and poultry, while grading is voluntary.

Vendor Selection. Selecting a vendor is an important part of the purchasing process. The foodservice manager must consider the extent to which the potential vendor will be able to meet the quality, service and cost expectations of the foodservice operations.

Figure 2.1 ■ USDA Inspection Stamps and Grade Shields

The relationship between the foodservice operator and the vendor must be one of mutual cooperation and trust. Past experience, good or bad, can serve as a basis for selection. Visits to the vendor's distribution center and inspections of delivery vehicles are advisable in the process of selecting a vendor. Once a reputable vendor is found, it is important to monitor quality continuously to ensure that the vendor has a long-term commitment to providing safe, quality food. Note that food prepared in the home cannot be served in quantity foodservice operations. Some of the considerations leading to selection of a safe food vendor are summarized as follows:

■ Is the vendor inspected by an independent source to ensure food safety?

■ Can the supplier provide you with written proof of government inspected meats?

■ Does the vendor have a Hazard Analysis Critical Control Point (HACCP) system in place?

■ Will the vendor allow the foodservice operator to set the delivery time?

■ Are the vendor's delivery vehicles generally clean and well maintained?

■ Are the delivery trucks for refrigerated foods refrigerated?

■ If purchasing frozen foods: do delivery trucks have freezer sections?

■ Can the vendor provide business references?

■ Can the vendor meet the delivery needs (daily, weekly, monthly) of the facility to ensure a safe flow of food?

■ Does the vendor have a reputation for providing quality products?

■ Will the vendor allow staff to inspect products upon receipt?

■ Is the vendor cooperative if you refuse products because of food safety concerns?

■ Where is the vendor located relative to the foodservice operation? A closer vendor would reduce delivery time and may reduce the possibility of contamination or time/temperature abuse of food.

Receiving

Receiving is an important phase in the flow of food through a foodservice operation. Food items must be carefully inspected before being placed into storage. The person receiving food must be knowledgeable of food safety, quality standards, and purchasing specifications. It is essential that foodservice operators inspect food immediately upon delivery. Food not meeting the operation's quality standards or of questionable food safety

should be refused and returned to the vendor. Schedule deliveries at times convenient for the receiver to inspect food carefully. Even government inspected foods must be examined, since Federal inspection programs do not inspect food at every step in the flow of food from producer to vendor to foodservice operation. Storage areas should be prepared to accept new deliveries and staff must be able to immediately date and store the food.

Potentially hazardous food must be received at a temperature of 41° F (5° C) or below. A frozen food must be received in a frozen state and at 0° F (-18° C) in most cases. Carefully inspect refrigerated and frozen foods for any sign of temperature abuse during transportation and delivery. For instance, large ice crystals are evidence that frozen food was thawed and refrozen — and should be rejected at delivery. When receiving food:

■ Check the temperature of refrigerated or frozen food.

Figure 2.2 ■ Receiving Checklist

Received by: _____ Date: _____

Page No. _____ of _____

Item	Actual Temp. ▲	Packaging Intact?		Use-by date valid?		Accepted ✔	Rejected✚ ✔	Stored✱ ✔
		Y✔	N✔	Y✔	N✔			

▲ Receive refrigerated food at 41° F (5° F) or below. Receive Frozen Food at 0° F (-18° C) Ice Cream may be received at 6° F - 10° F (-14.5° C to -12° C)

✚ If rejected, write comments on back of this form

✱ Check to confirm that the item was stored

Figure 2.3 ■ Verifying Food Temperature During Receiving

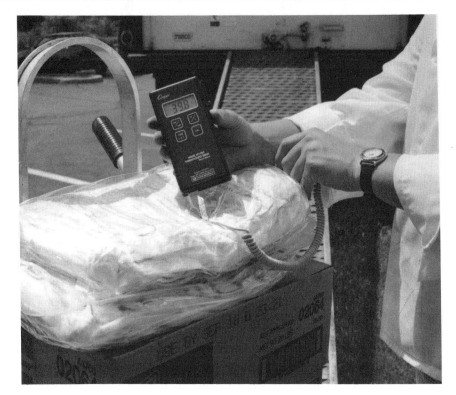

Courtesy: Cooper Instrument Corporation

■ Don't accept food in damaged packaging.

■ Check for inspection stamps and date codes.

■ Verify freshness by color, odor, touch and package condition.

■ Look for obvious signs of pest infestation and/or spoilage.

Modified Atmosphere Packaged Foods. Recent innovations in packaging include **MAP** (modified atmosphere packaged) and **CAP** (controlled atmosphere packaged) foods.

Figure 2.4 ■ Quality Indicators for Receiving Safe Food

Food Item	Accept...	Reject...
Fresh Beef	• light pink to bright, red color (aged beef may be darker) • firm and elastic	• dark brown or greenish color • sour or rancid smell
Fresh Poultry	• slightly yellow appearance • firm flesh • not sticky to touch	• dark appearance under wings • sticky to touch
Marine Foods	• gills of fresh fish are pink • no iridescence (shimmering) • eyes clear, not sunken	• excessive fish or ammonia odor
Milk and Dairy Products	• intact packaging • clean containers • butter with firm texture • cheese free of mold	• damaged or leaking containers • expired dates
Eggs and Egg Products	• clean eggs, uncracked shells	• cracked or dirty eggs
Fresh Produce	• bright color, no mold or wilt	• signs of insect damage or plant disease • bruises or soft produce
Canned Foods	• no dents • all foods labelled	• dented or rusted cans
Dry Goods	• intact packaging	• ripped or torn packaging

These types of packaging extend freshness by maintaining a reduced-oxygen environment. Although the packaging can extend shelf life, it will not prevent microbial growth if held at an unsafe temperature. When receiving MAP or CAP foods:

■ Read the label. Recommended handling practices vary from product to product.

■ Check for a *time/temperature indicator* (TTI) strip. This strip changes color if the product is outdated or has been temperature abused.

■ Check for an expiration date on the package.

■ Check for air bubbles; vacuum packaged foods should be free of air bubbles.

Ultra-Pasteurized Processed Foods. One type of food processing that destroys harmful microorganisms is called *ultra-pasteurization*. In this process, food is treated with high temperatures for a short time. Ultra-pasteurized foods require refrigeration. Ultra-pasteurized foods that have also been aseptically packaged are labeled **UHT**. Some varieties of coffee creamers are examples of foods packaged in this manner. UHT packaged foods may be received and stored unrefrigerated until after opening.

Storage

Proper storage is an important aspect of food protection. Storage is more complex than simply having adequate space. Storage involves the variables of light, temperature, ventilation, and air circulation. Food must be stored 6" (15 cm) above the floor, on clean racks or shelves, and in a manner that prevents environmental or cross contamination. Food not intended for further preparation before serving must be stored in a way that prevents contamination from food requiring preparation. For instance, raw foods — such as meat — should never be

Figure 2.5 ■ Storage Temperatures

Dry Storage	50˚ F - 70 ˚F (10 ˚C - 21 ˚C)
Refrigerated Storage	32˚ F - 40 ˚F (0˚ C - 5˚ C)
Deep Chilling Storage	26˚ F - 32˚ F (-3˚ C - 0˚ C)
Freezer Storage	-10˚ F to 0˚ F (-23.5˚ C to -18 ˚C)

stored above ready-to-eat foods — such as fresh fruit, salads, or desserts. The raw food may splash or drip onto the ready-to-eat food and result in cross contamination. Food may not be stored in locker rooms, toilet areas, dressing rooms, in garbage rooms, mechanical rooms, under sewer lines that are not adequately shielded, under open stairwells, under a water leak, or near any other potential source of contamination.

Every food has a shelf life, which is the amount of time that the product can reasonably be expected to remain safe if stored properly. Once food is received, it should be date labeled. Foodservice personnel must monitor the dates on foods to ensure they are utilized before the expiration of their shelf lives. New deliveries must be stored behind previous deliveries to ensure that the oldest product is used first. This method of stock rotation is called *First In, First Out* (FIFO).

Containers holding food or food ingredients that are removed from their original package for use in the foodservice operation should be labeled with the common name of the food and the date the package was

Figure 2.6 ■ Storage Temperature Log

Storage Temperature Log

Storage Area:_____

Page No. _____ of _____

Date	Storage Temperature		Comments/Corrective Actions
	A.M.	P. M.	

Dry Storage: 50˚ F - 70 ˚F (10˚ C to 21˚ C)   Refrigerated Storage: 32˚ F - 40˚ F (0˚ C - 5˚ C)

Deep Chilling Storage: 26˚ F - 32˚ F (-3˚ C to 0˚ C)   Freezer Storage: -10˚ F to 0˚ F (-23.5˚ C to -18˚ C)

Figure 2.7 ■ Guidelines for Purchasing, Receiving and Storing Food

	Purchase	Receive	Store	Shelf Life
Meat	USDA or state inspected	Receive frozen at 0° F (18° C). Receive refrigerated at 41° F (5° C) or below.	Store frozen at 0° F (-18° C) or below. Store refrigerated at 32° F - 41° F (0° C - 5° C).	Fresh beef: 6-12 mos. in freezer 3-6 days in refrigerator Ground beef: 3-4 months in freezer 1-2 days in refrigerator
Poultry	USDA or state inspected	Receive frozen at 0° F (18° C). Receive refrigerated at 41° F (5° C) or below.	Store frozen at 0° F (-18° C) or below. Store refrigerated at 32° F - 41° F (0° C - 5° F).	6 months in freezer 1-2 days in refrigerator
Marine Foods	From an approved supplier; shellfish must have shell-stock identification tag	Receive frozen at 0° F (18° C). Receive refrigerated at 41° F (5° C) or below.	Store frozen at 0° F (-18° C) or below. Store refrigerated at 32° F - 41° F (0° C - 5° F).	All: 2- 4 months in freezer 1-2 days in refrigerator
Eggs and Egg Products	Grade AA or A No more than a 2-week supply of fresh eggs	Receive frozen egg products at 0° F (18° C). Receive refrigerated at 41° F (5° C) or below.	Store frozen at 0° F (-18° C) or below. Store refrigerated at 32° F - 41° F (0° C - 5° F).	Carton eggs: follow label directions Fresh eggs: no more than 2 weeks in refrigerator
Milk and Dairy Products	Pasteurized milk and milk products	Receive frozen (ice cream) at 6° F - 10° F (-14.5° C - 12° C) Receive refrigerated at 41° F (5° C) or UHT processed may be stored unrefrigerated.	Store frozen at 0° F (-18° C) or below. Store refrigerated at 32° F - 41° F (0° C - 5° F). Unopened UHT package may be stored above 41° F.	Milk: use by expiration date on carton Ice Cream: 6 months in freezer

opened. Slotted shelving allows for proper air circulation around food, and should not be covered with foil or other material.

Food storage areas must be equipped with thermometers accurate to ±3° F (1.5° F). The temperature of storage areas should be monitored not less than once a day and the actual temperatures documented on a temperature log.

Dry storage area temperatures should be maintained between 50° F - 70° F (10° C - 21° C). For short-term holding of perishable and potentially hazardous foods, maintain refrigerators between 32° F - 40° F (0° C - 5° C). Lower temperatures may damage food and affect quality for some items such as produce. Ideally, meats, poultry and fish should be stored in a deep chilling unit. Maintaining refrigeration temperatures below 38° F (2° C) will ensure that all food product temperatures are in the safe zone for cold food temperatures 41° F (5° C) or below.

Key Concepts

■ Purchase food only from an approved source.

■ Inspect food upon receipt for quality and temperature standards.

■ Store food immediately upon receipt.

■ Maintain clean storage areas at the appropriate temperature.

■ Date mark all deliveries and rotate supplies using the FIFO principle.

■ Monitor food temperatures during the phases of receiving and storage.

■ Protect food from cross contamination during storage.

■ Conduct regular self-inspections to ensure that food is being stored safely.

Case Study

While randomly checking the temperatures of a delivery, you discover that the temperature of fresh beef steaks is 55° F (13° C). You step into the truck to explain the problem to the driver and notice that the refrigerated truck feels warm too.

☛ 1. What is a safe temperature for receiving fresh beef steaks?

☛ 2. What is an appropriate corrective action to take?

☛ 3. How could you verify the safety of other foods arriving on the same order?

Review Questions

1. USDA meat grades indicate information concerning a meat's:

    A. quantity

    B. quality

    C. packing date

    D. sanitary condition

2. A foodservice manager finds that individually quick frozen chicken pieces are frozen together upon delivery. The manager should:

    A. return the chicken to the supplier

    B. thaw the chicken in the refrigerator before cooking

    C. thaw the chicken under running water before cooking

    D. separate the chicken with a meat mallet before cooking

3. Raw poultry is acceptable for delivery if it has a maximum internal temperature of:

    A. 60° F (15.5° C)

    B. 55° F (13° C)

    C. 45° F (7° C)

    D. 41° F (5° C)

4. Meat should be inspected for safety and wholesomeness when delivered because Federal and state inspection programs:

    A. mandate inspection of all food products in the United States

    B. apply only to commodity products distributed by the government

    C. fail to check food products at every point prior to delivery

    D. encourage voluntary inspection of meat

5. Upon delivery, fresh meat should appear:

   A. slightly gray with some brown coloration

   B. bright, cherry red

   C. dark brown; some green discoloration is acceptable

   D. dark brown with no discoloration

6. Identification tags for shellfish should be kept on file for:

   A. 30 days

   B. 60 days

   C. 90 days

   D. 120 days

7. An appropriate temperature range for a dry storage area is:

   A. 30° F - 50° F (1° C - 10° C)

   B. 40° F - 60° F (4.5° C - 15.5° C)

   C. 50° F - 70° F (10° C - 21° C)

   D. 60° F - 80° F (15.5° C - 26.5° C)

8. Food storage areas should be equipped with thermometers accurate to:

   A. ±1° F

   B. ±2° F

   C. ±3° F

   D. ±4° F

9. Food should be stored at least:

    A. 2" off of the floor

    B. 4" off of the floor

    C. 6" off of the floor

    D. 8" off of the floor

10. Covering storage shelves with aluminum foil or other material:

    A. maintains temperature

    B. facilitates cleaning

    C. prevents proper air flow

    D. reduces pest infestations

Preparation and Service of Safe Food

Chapter Objectives

■ Describe how food can be cross contaminated

■ Identify safe methods for thawing food

■ List safe temperature ranges for the endpoint of cooking

■ Describe how to hold hot and cold foods properly

■ Identify safe methods for cooling food

■ Identify food protection practices for self-service areas

■ Describe the proper way to reheat food

The movement of food through the various stages of receiving, storing, preparing, serving, and reheating has been described as the flow of food. Each step in this flow presents new risks and new challenges for food protection.

Preparing Safe Food

Temperature is a critical variable in protecting food. Food temperature must be monitored at each step in the flow of food: purchasing, receiving, storing, preparing, serving, transporting, cooling and reheating. Maintain food out of the hazard zone for food temperatures. To ensure that food is being safely maintained, keep temperature logs in all storage, preparation, and service areas. Involve all staff members in monitoring food temperatures, and be sure all staff are trained in proper procedures for recording food temperatures.

Time is also an important control technique for food protection. When moving food from storage to preparation areas, label it to indicate the latest time for use. Once food has been removed from a monitored, temperature-controlled environment, serve or dispose of it within four hours. Dispose of food if it is in unmarked containers, or if its expiration date has passed. Communicate food labeling

and monitoring practices to all members of the foodservice team.

Cross Contamination. During all phases of food preparation and service, foodservice professionals must take steps to prevent cross contamination. Cross contamination is the transfer of pathogens from one food to another food by contact with a non-food item, such as a work surface or the hands of a foodservice worker. Cross contamination can also result from food spills or splashes. To prevent cross contamination:

■ Follow proper procedures for cleaning and sanitizing equipment and utensils.

■ Observe good hand washing practices.

■ Store raw foods below ready-to-eat or prepared foods.

Raw fruits and vegetables must be thoroughly washed in water to remove soil and other contaminants. They should be washed before being cut, combined with other ingredients, cooked, served, or offered for service in a ready-to-eat form except when intended for washing by the client.

Thawing. Foods must be thawed in a manner that avoids placing the food in the hazard zone for food temperatures for an excessive amount of time. Frozen food should never be thawed at room temperature. Following are four options for thawing food safely:

■ Thaw gradually under refrigeration that maintains the food temperature at 41° F (5° C) or less.

■ Thaw completely submerged under cold 70° F (21° C) or less, potable running water with water pressure sufficient to continuously agitate any food particles off the surface. With this method, thawing time should be less than two hours, or until food reaches 41° F (5° C). This method may not be effective for thawing large pieces of meat.

■ Thaw during the cooking process by cooking frozen food in the oven in a continuous process.

■ Thaw and cook single-service completely in the microwave from a frozen state.

Food that is removed from the freezer to gradually thaw under refrigeration should be date-labeled to indicate the date by which the food should be consumed. The food should be consumed seven days or less after the food was removed from the freezer and should be held at 41° F (5° C) or below.

Cooking. Cooking is a very important step in the flow of food. Failure to cook food thoroughly is a frequent cause of foodborne illness. Cooking food thoroughly destroys most biological hazards. Pork and game animals need a higher cooking temperature, due to the possible presence of parasites. Ground beef and pork require a higher cooking temperature due to the increased surface area resulting from the meat grinding process. The key concept for cooking is to cook food quickly and thoroughly. "Thoroughly" means to a recommend internal temperature at the endpoint of cooking. Cooking temperatures for foods should be measured with a calibrated measuring device, such as a thermometer or thermocouple. Temperature should be measured in the part of the food that is heated last — usually the center or thickest part of the food. Because cooking is such a significant step, it is

Figure 3.1 ■ Cooking Time and Temperature Guide			
Food Item	**Temperature**	**Time**	**Comments**
Beef Roasts	145° F (63° C)	for 3 minutes	
	140° F (60° C)	for 12 minutes	
	130° F (54° C)	for 121 minutes	
Ground Beef	155° F (68° C)	for 15 seconds	
Pork	155° F (68° C)	for 15 seconds	
Poultry	165° F (74° C)	for 15 seconds	
Seafood and Shellfish	155° F (68° C)	for 15 seconds	
Eggs and Egg Products	155° F (68° C)	for 15 seconds	held on steam table
	145° F (63° C)	for 15 seconds	cooked to order
Stuffed Food	165° F (74° C)	for 15 seconds	

Figure 3.2 ■ Cooking Temperature Log

Cooking Temperature Log

Date:_____   Page No. _____of _____

Item	Time	Temperature	Cook's Initials	Endpoint ✔ ▼

▼ Check to indicate that this was the endpoint of cooking

often referred to as a critical control point. Critical control points are explained in Chapter 4.

Microwave Cooking. When preparing food in a microwave:

■ Rotate or stir the food item throughout or at the midpoint of cooking to distribute heat evenly.

■ Cover to retain surface moisture.

■ Cook to a temperature of at least 165°F (74°C) in all parts of the food item.

■ Allow to stand covered for two minutes before serving to obtain temperature equilibrium.

Serving Safe Food

During service, a combination of time and temperature is used to prevent or slow microbial growth. Potentially hazardous foods must be held out of the hazard zone for food temperatures.

Hot Holding. Hot foods must be held at 140° F (60° C) or above, with the exception of roasts cooked to a time and temperature shown in Figure 3.1. Stir foods during holding to redistribute heat throughout the food product. Keep food containers covered to retain heat and to prevent environmental contaminants from entering the food.

Steam tables and food warmers are examples of equipment that may be used to retain heat and control food temperatures. Food warmers are usually equipped with an ambient thermometer. Steam tables usually have a temperature control knob but may not measure the actual water temperature. When holding hot foods for service, measure the food temperature at least every two hours. If the food temperature falls into an unsafe range, immediately follow procedures for reheating previously cooked food.

Cold Holding. Cold food must be held at 41° F (5° C) or less. Holding food below this temperature slows the growth of bacteria. Special considerations are required for using ice as a

Figure 3.3 ■ Temperatures for Preparing and Serving Food

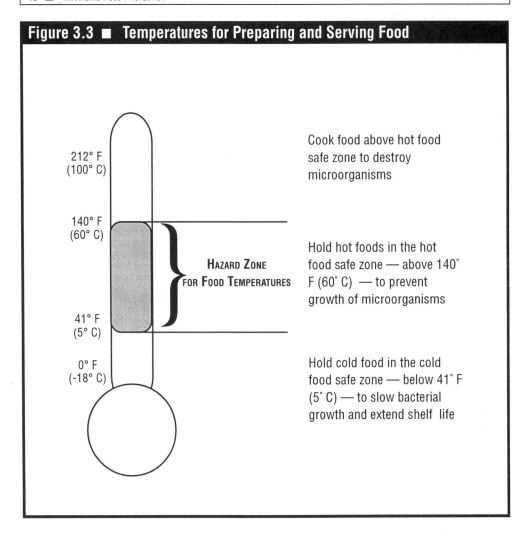

212° F
(100° C)

Cook food above hot food safe zone to destroy microorganisms

140° F
(60° C)

HAZARD ZONE
FOR FOOD TEMPERATURES

Hold hot foods in the hot food safe zone — above 140° F (60° C) — to prevent growth of microorganisms

41° F
(5° C)

0° F
(-18° C)

Hold cold food in the cold food safe zone — below 41° F (5° C) — to slow bacterial growth and extend shelf life

coolant. Packaged food may not be stored in direct contact with ice or water if the entry of water is possible due to the nature of the packaging or its positioning in the ice or water. Ice that has been used as a coolant may not be used as a food or food ingredient. Unpackaged food may not be stored in direct contact with undrained ice, with these exceptions:

■ Whole, raw fruits or vegetables — such as celery or carrot sticks or whole potatoes — may be immersed in ice or water.

■ Raw chicken and raw fish that are received immersed in ice in shipping containers may remain in that condition while in storage awaiting preparation or service.

Figure 3.4 ■ Monitoring Temperatures During Preparation and Service

Courtesy: Cooper Instrument Corporation

Figure 3.5 ■ Holding Temperature Log

Holding Temperature Log

Date:_____   Page No. _____of _____

Item	Time	Temperature	Cook's Initials	Corrective Action

Hold cold foods below 41° F (5° C) ✍ **Hazard Zone** ☞ Hold hot foods above 140° F (60° C)

During holding and service, utensils should be kept in the food. This is especially important with moist foods such as ice cream or mashed potatoes. Utensil handles should be kept above the top of the food and the container, or in running water with sufficient velocity to remove food particles continuously. For serving a food that is not potentially hazardous, the serving utensil may be held in a clean, protected container. During service, food on display must be protected from contamination. During transport, food items must continue to be maintained at a safe temperature, and food must be covered to prevent contamination.

Self-Service Food Bars. Self-service food bars, such as buffets and salad bars, present special food protection risks. Customer actions and habits create the potential for food contamination. Label all food items in self-service areas to discourage customers from sampling food items. Sneeze guards must be used on service lines,

salad bars and display cases. Self-service food bars should be monitored by foodservice workers trained in food protection practices. Clients should be required to use a clean plate for each service from a food bar to prevent contamination from soiled dishes and utensils.

Cooling. After proper cooking, potentially hazardous foods must be cooled as rapidly as possible. When foods begin to cool, they enter the hazard zone for food temperatures. The foodservice professional must take steps to reduce the time the food spends in the hazard zone for food temperatures. According to the FDA Food Code, foods should be cooled from $140°$ F $(60°$ C) to $70°$ F $(21°$ C) within two hours, and from $70°$ F $(21°$ C) to $41°$ F $(5°$ C) within an additional four hours. To facilitate cooling in a walk-in cooler, it is acceptable to leave foods loosely covered or uncovered if protected from overhead contamination. Once cooled, food should be covered, dated, and labeled.

Figure 3.6 ■ Proper Cooling Methods

■ Placing food in a shallow container (2" deep) to increase surface area and reduce cooling time

■ Dividing a large food mass into several smaller masses (e.g. a 20-lb. roast into 5 4-lb. pieces)

■ Stirring the food in a container placed in an ice water bath

■ Using rapid cooling equipment, such as a blast chiller

■ Adding ice as an ingredient

Figure 3.7 ■ Cooling Food with the Ice-Bath Method

Courtesy: Cooper Instrument Corporation

Storing Prepared Foods. Cooked foods should not be left unrefrigerated for more than four hours. Put leftovers in small, clean containers and refrigerate or freeze immediately. Potentially hazardous food prepared in the foodservice operation and held at or below 41°F (5°C) should be labeled at the time of preparation with the date by which the food should be consumed. Prepared foods should be used within seven days or less of preparation if held at 41°F (5°C).

Reheating. Frozen, ready-to-eat food taken from a commercially prepared, intact package must be heated to at least 140°F (60°C) for hot holding. Reheating for hot holding should be done as rapidly as possible.

A previously cooked food (a food that was prepared, cooked, and cooled in the foodservice operation) must be reheated to an internal temperature of 165°F (74°C) within two hours. A previously cooked food must be reheated to attain a temperature of 165°F (74°C) for a minimum of 15 seconds.

Key Concepts

■ Most foodborne illnesses are the result of mistakes or mishandling in the final stages of preparation.

■ Temperature is a critical variable in protecting food.

■ Time is also an important control technique for food protection.

■ Cross contamination is the transfer of pathogens from one food to another food by contact with a non-food item such as a work surface or the hands of a foodservice worker.

■ Foods must be thawed in a manner that avoids placing the food in the hazard zone for food temperatures for an excessive amount of time.

■ Failure to cook food thoroughly is a frequent cause of foodborne illness.

■ The key concept for cooking is to cook food quickly and to cook food thoroughly.

■ Hot foods must be held at 140° F (60°C) or above.

■ Cold food must be held at 41°F (5°C) or less.

■ A previously cooked food (a food that was prepared, cooked, and cooled in the foodservice operation) must be reheated to an internal temperature of 165°F (74°C) within two hours.

Case Study

A caterer prepared food for 300 guests, but only 100 guests were in attendance. Pat, the cook, must store 20 gallons of leftover tomato meat sauce. Pat removes the sauce from the food warmer where it was being held. She divides the sauce into two 8" deep stainless steel pans and tightly covers the pans with plastic wrap. Pat labels the food with the date and time, and then immediately stores the leftover sauce in the cooler.

☞ 1. What did Pat do correctly?

☞ 2. What did Pat do incorrectly?

☞ 3. What are the methods that Pat could have used to cool the leftover sauce safely?

Review Questions

1. Which of the following is the most acceptable and safe way to thaw a frozen 20-lb. turkey?

    A. at room temperature

    B. under cold running water

    C. in the microwave

    D. in a bain marie

2. The minimum safe internal temperature for cooking ground beef is 15 seconds at:

    A. 140° F (60° C)

    B. 145° F (63° C)

    C. 155° F (68° C)

    D. 165° F (74° C)

3. Stuffed pork chops should be cooked to a minimum internal temperature of:

    A. 140° F (60° C) for 15 seconds

    B. 145° F (63° C) for 15 seconds

    C. 155° F (68° C) for 15 seconds

    D. 165° F (74° C) for 15 seconds

4. When preparing food in the microwave, cook to a minimum endpoint temperature of:

    A. 140° F

    B. 150° F

    C. 155° F

    D. 165° F

5. A large quantity of food can best be cooled by:

    A. placing in the freezer

    B. stirring frequently in an ice bath

    C. cooling at room temperature until it reaches 70° F (21° C)

    D. placing in a deep container in a walk-in cooler

6. To prevent cross contamination:

    A. sanitize equipment after each use during preparation

    B. rinse utensils with warm water after each use

    C. store raw foods above ready-to-eat foods

    D. store meat and poultry in separate coolers

7. Which of the following is NOT a good practice for self-service food bars?

    A. label all food items for customers

    B. use a sneeze guard

    C. encourage customers to re-use their own plates

    D. have trained staff monitor the bar frequently

8. When holding hot foods, monitor temperature at least every:

    A. hour

    B. 2 hours

    C. 3 hours

    D. 4 hours

9. Reheat all previously cooked foods to a minimum internal temperature of at least 165° F (74° C) for 15 seconds within:

   A. 1 hour

   B. 2 hours

   C. 3 hours

   D. 4 hours

10. According to the FDA Food Code, foods should be cooled from 140° F (60° C) to 70° F (21° C) within 2 hours, and from 70° F (21° C) to 41° F (5° C) within an additional:

   A. hour

   B. 2 hours

   C. 4 hours

   D. 6 hours

The Hazard Analysis Critical Control Point System

Chapter Objectives

■ Describe the development of HACCP food safety systems

■ Define common HACCP terms

■ List the seven principles of HACCP

■ Describe the steps in implementing a HACCP system

Hazard Analysis Critical Control Point (HACCP) is an important development in food protection. The HACCP system is a preventive food safety program that greatly reduces the likelihood of foodborne illness. The Hazard Analysis Critical Control Point system or HACCP (pronounced Has-sip) is a food safety system developed 30 years ago to ensure a safe food supply for astronauts. It is now being implemented in the food processing and service industries. The most basic concept underlying HACCP is that of prevention. Essentially, HACCP involves identifying where and how food could become unsafe (hazard analysis) and establishing standard operating procedures to protect food (critical control points).

The advantages of a HACCP system include:

■ It focuses on preventing hazards from contaminating food

■ It applies scientific principles to food handling and protection practices

■ It develops universal practices that can be applied in various foodservice operations and segments of the foodservice industry

■ It provides a record for regulatory agencies that describe food handling practices over time, rather than just on the day of inspection.

The success of HACCP is dependent on a number of variables. Foodservice managers learn HACCP principles and concepts and demonstrate their ability to effectively organize a HACCP system by applying their knowledge. Foodservice managers must educate all members of their foodservice team on the principles of HACCP and on the team's role in its implementation. Staff training is essential to the successful implementation of HACCP.

Implementing a HACCP System

1- **Analyze Hazards.** Identify actual and potentially hazardous foods associated with your menu. These items could be hazardous due to ingredients, the processes involved in preparation, how the product is held or handled during service, or its ultimate use. Hazards may vary from one foodservice operation to another, even with the production of the same menu item, because of differences in ingredient sources, formulations, production equipment, preparation methods, storage and experience or knowledge of foodservice personnel.

While HACCP is concerned with the safety of all food, particular attention should be given to potentially hazardous foods. Potentially

Figure 4.1 ■ Common HACCP Terms

Hazard is unacceptable contamination, unacceptable growth, or unacceptable survival of harmful microorganisms. The major types of hazards are biological, chemical, and physical. Hazards are identified by following food through the flow of food, including each step from receiving through service to the client.

Severity is the significance of the hazard or the degree of consequences that can result when a hazard exists.

Monitoring is the checking that a processing or handling procedure does not exceed the established critical limit at each critical control point. It involves systematic observation, measurement and/or recording for prevention or is an estimate of the probability of occurrence of a hazard or several hazards. More than one observation may be necessary at a particular critical control point. The monitoring procedures chosen must enable action to be taken to correct an out-of-control situation or to bring the product back into acceptable limits.

Critical control point (CCP) is an operation or step of a operation, at or by which a preventive measure can be exercised that will eliminate, prevent or minimize a hazard that occurred prior to that point.

Critical limits are specified limits or characteristics of a physical, chemical or biological nature.

Verification involves the use of equipment to determine that the HACCP system is in place and achieving the desired objectives.

hazardous foods are usually high-protein foods such meat, poultry, fish, seafood, eggs, milk or dairy products. These foods can support rapid bacterial growth. Estimate the risks involved and your foodservice operation's ability to control these risks. *Risks* refer to the chance that a condition or set of conditions will lead to a hazard. Risks vary based on the size and type of operation, type of customers, complexity of recipes, number of potentially hazardous menu items and ingredients, and employee skill level. Determine the likelihood of incoming foods introducing foodborne pathogens into the establishment. Evaluate storage to determine if it is adequate to maintain appropriate temperatures. Determine whether potentially hazardous foods are being stored beyond the storage capacity of the facility.

2- Identify Critical Control Points.

These are points in the flow of food at which a potential hazard can be controlled or eliminated. Examples are cooking, cooling and reheating. A *critical control point* (CCP) is any point or step in the preparation of a specific menu item where some type of action must be taken to prevent or minimize the risk of foodborne illness. Critical control points vary depending on the food item being prepared and the preparation techniques being used. Inspecting food upon receipt for package integrity is a control point and suggested practice, but it is not a critical control point because no specific action is being taken to destroy or slow the growth of harmful bacteria.

By comparison, cooling food is a critical control point. Failure to take a specific action — proper cooling technique — can result in the growth of pathogens and a risk to consumers of the food product. Most critical control points are present during the steps of preparation and service. When determining critical control points, ask these questions:

■ When could contaminants enter the food item? Consider storage,

Figure 4.2 ■ Seven Principles of HACCP

1. Analyze Hazards

Potential hazards associated with a food are identified. The hazard could be biological (bacteria), chemical (a cleaning solution) or physical (metal shavings).

2. Identify Critical Control Points

These are the points in the flow of food at which a potential hazard can be controlled or eliminated. Examples are thawing, cooking, cooling and reheating.

3. Establish Critical Limits for Control Points

For a cooked food, this might include setting a minimum internal temperature for the endpoint of cooking, or the amount of time allowed to cool a food to a specified temperature.

4. Establish Procedures for Monitoring Control Points

Such procedures might include determining how and by whom cooking time and temperature will be monitored.

5. Establish Corrective Actions

When monitoring shows that a critical limit has not been met, there must be an established procedure to follow. For example: Receiving refrigerated food is a control point. The critical limit is to receive at 41° F (5° C). If refrigerated food arrives at 50° F, the corrective action would be to reject the delivery.

6. Establish a Record-Keeping System

A record-keeping system documents HACCP activities and includes time and temperature monitoring records.

7. Establish Procedures to Verify that the System is Working

Calibration of temperature monitoring devices and review of HACCP logs for accuracy and completeness will help to verify the system is fulfilling its mission to improve food safety.

preparation, and service of the food item. Consider physical and chemical contaminants as well as biological.

■ What steps will be taken to limit the growth or destroy harmful pathogens? What storage tem-

perature will the food require? Will it be thawed, cooked, cooled or reheated?

■ How much time will this food item spend in the temperature hazard zone?

Figure 4.3 ■ Critical Control Points in the Flow of Food

Processing Step	Control Points	CCP?
Purchase	Approved Supplier	No
Receive	Receive at acceptable temperature	No
Store	Store immediately at acceptable temperature	No
Prepare	Thaw with an approved practice	No
Cook	Cook to a safe internal temperature	Yes
Hold	Hold at a safe temperature	Yes
Serve	Maintain at a safe temperature during service	Yes
Cool	Cool with an approved method	Yes
Store	Refrigerate or freeze Keep covered Use within 7 days or less if highly perishable	No No No
Reheat	Reheat to a safe temperature	Yes

Figure 4.4 ■ Examples of Critical Limits

CCP	Critical Limit
Cooking	Cook to an internal temperature of 165°F (74°C)
Holding	Hold at 140°F (60°C)
Cooling	Cool from 140°F (60°C) to 70°F (21°C) in 2 hours, and from 70°F (21°C) to 41°F (5°C) within an additional 4 hours
Reheating	Reheat to 165°F (74°C) in two hours or less

3- Establish Critical Limits for Control Points. There must be a critical limit or procedure for each CCP identified. There may be more than one control for each CCP. Critical limits are observable and measurable. For a cooked food, for example, this might include setting the minimum internal temperature for the endpoint of cooking, or the amount of time allowed to cool a food.

4- Establish Procedures for Monitoring Control Points. Follow ingredients as they flow through a typical recipe. Include the flow from receiving, through preparation, service, cooling, storage, and reheating. Focus on CCPs through the flow. Determine how and by whom cooking time and temperature will be monitored. Monitoring procedures may be included in standardized recipes. Observe and measure to determine whether critical limits are met. Identify deviations from standard operating procedures. Critical limits may also be included in recipes, and foodservice personnel should be required to measure all CCPs and document in HACCP logs.

Figure 4.5 ■ Identifying Potentially Hazardous Foods on a Menu

MENU

Breakfast   **Orange Juice**

                        (**Scrambled Eggs**)

                        (**Canadian Bacon**)

                        Toast with Jelly

Lunch   **Tossed Salad**

                        (**Beef Barley Soup**)

                        (**Turkey Club Sandwich**)

                        Cherry Cake

Dinner   **Peach Salad**

                        (**Stuffed Pepper**)

                        (**Whipped Potatoes**)

                        Baby Carrots

                        (**New York Style Cheesecake**)

5- Establish Corrective Actions. When monitoring shows that a critical limit has not been met, there must be an established procedure to follow. Corrective actions are specific to each foodservice operation. Corrective actions should be clearly written and understood by foodservice personnel.

For example, receiving refrigerated food is a control point. The critical limit is to receive at 41° F (5° C). If refrigerated food arrives at 60° F (15.5° C), the corrective action would be to reject the delivery. Other examples of corrective actions might be: contacting the foodservice manager for direction, reheating at once to a specific temperature, or disposing of food that was improperly cooled.

6- Establish a Record-Keeping System. A record-keeping system documents HACCP activities and includes time and temperature monitoring records. Use forms to document the flow of food product, time and temperature, and corrective actions. Figures 3.2 and 3.5 in Chapter 3 are examples of HACCP logs for holding and cooking. Maintain daily HACCP logs in a notebook accessible to all members of your foodservice team.

7- Establish Procedures to Verify that the System is Working. Calibration of temperature monitoring devices and review of HACCP logs for accuracy and completeness will help to verify the system is fulfilling its mission to improve food safety. Review HACCP logs, records of monitoring CCPs and corrective actions taken by staff members. Conduct random checks to ensure that CCPs are being monitored appropriately by staff and that temperature monitoring equipment is working properly. Managers should be aware of **dry lab**, the practice of entering time and temperature measurements without actually taking them. Dry lab can undermine the success of any HACCP program. Update your HACCP system as changes in suppliers, menu, and preparation methods are made.

A HACCP Plan Case Example

A **HACCP plan** is a written document that describes the formal procedures for following HACCP principles. In this example, Community Hospital is implementing a HACCP plan. To begin the process, Community Hospital has assembled a HACCP team that includes the: Dietary Manager,

Figure 4.6 ■ Flow Chart for Stuffed Pepper Recipe

Menu Item: Stuffed Peppers   Yield: 24 portions

Ingredients	Amount
Green peppers, large	12
Rice	10 oz
Ground beef	2 1/2 lb
Onion, fine dice	10 oz
Chopped garlic	1 tsp
Tomato sauce	3 pt
Oil	3 oz
Water	

Purchase
Approved source •
USDA inspected
ground beef

Receive
All packaging intact •
Beef received at 0°F
(-12°C) or below

Storage
All packaging intact •
Beef stored at 0°F
(-12°C) or below

Pre-Preparation
Rice cooked and cooled
properly • Pepper
blanched • Onions diced •
Beef thawed in refrigerator
on bottom shelf 2 days
prior to prep

Prepare
Mix all ingredients
together • Place in
cooler until time to
cook

Cook
Bake at 350°F (175°C)
• Cook to 165°F
(74°C) for 15 seconds
• CCP

Holding
Place on steam table and
cover • Check temperature
every hr • Hold above
140°F (60°C) during
service • CCP

Cooling/Storage
Cool from 140°F (60°C) to
70°F (21°C) within 2 hrs,
and from 70°F to 41°F
(5°C) within 4 more hrs •
Store at 41°F (5°C) or
below — top shelf, 3 days
or less • CCP

Reheat
Reheat in 350°F
(175°C) oven • Reheat
to 165°F
(74°C) within 2 hrs •
CCP

Figure 4.7 ■ HACCP Plan for Stuffed Peppers

STEPS 1-2: ANALYZE HAZARDS, IDENTIFY CRITICAL CONTROL POINTS

STEP		CCP
Purchase	USDA inspected beef	No
Receive	Frozen beef at 0˚ F (-18˚ C) or below	No
Store	0˚ F (-18˚ C) or below	No
Prepare	Thaw at 41° F (5° C) or below	No
Cook	Cook to 165° F (74° C)	Yes
Hold	Hold at 140° F (60° C)	Yes
Cool	Cool from 140˚ F (60˚ C) to 70˚ F (21˚ C) within 2 hrs, and from 70˚ F (21˚ C) to 41˚ F (5˚ C) within 4 hrs	Yes
Reheat	Reheat to 165˚ F (74˚ C) within 2 hrs	Yes

Figure 4.7 ■ HACCP Plan for Stuffed Peppers *continued*

STEPS 3-6: ESTABLISH CRITICAL LIMITS, PROCEDURES FOR MONITORING, CORRECTIVE ACTION, RECORD-KEEPING

CCP	CRITICAL LIMIT	STANDARD	CORRECTIVE ACTION
Cooking Beef/ Rice Mixture	Cook to 165° F (74° C) for at least 15 sec.	Monitor: observation Stuffed food: Cook thoroughly to destroy pathogens (high risk for E. coli). Monitor: record temperature	Continue cooking until critical limit is reached.
Holding and Serving Stuffed Pepper	Maintain tempera-ture at 140° F (60° C) or above.	Monitor temperature every 1/2 hr during service.; record on HACCP log. Monitor: record temperature	If temperature falls below 140° F (60° C), reheat immediately to 165° F (74° C).
Cooling	Cool from 140° F (60° C) to 70° F (21° C) within 2 hrs, and from 70° C to 41° F (5° C) within 4 more hrs. Store at 41° F (5° C) or below.	Monitor temperature every hr during cooling. Record temperature on HACCP cooling log. Monitor: record temperature	Dispose of food that is not properly cooled.
Reheating	Reheat to 165° F (74° C) within 2 hrs.	Monitor: Check time/temperature during reheating.	Dispose of food that is not properly reheated.

Receiving Clerk, Cook, and Trayline Supervisor.

1- Analyze Hazards. The HACCP team begins by identifying the potentially hazardous foods on each day of the menu cycle. The menu for the first day of the menu cycle appears in Figure 4.5. The potentially hazardous foods are circled.

2- Identify Critical Control Points. After identifying the potentially hazardous foods, the team begins to develop a flow chart to identify the critical control points in the flow of food for each potentially hazardous food. Flow charting requires taking a closer look at the recipe for potentially hazardous foods. The flow chart for the stuffed pepper recipe appears in Figure 4.6.

3- Establish Critical Limits for Control Points. After the identifying the critical control points, the team next determines the critical limit for each critical control point. The critical limits are described in Figure 4.7.

4- Establish Procedures for Monitoring Control Points. The team next determines who will be responsible for monitoring the critical limits. The team must decide what forms will be necessary to document monitoring activities, and how staff will be informed of their responsibilities for monitoring. Monitoring information is included in the HACCP Plan (Figure 4.7) and in the recipe (Figure 4.8).

5- Establish Corrective Actions. When monitoring shows that a critical limit has not been met, there must be an established procedure to follow. For example, cooling food is a CCP. The critical limit is to cool from 140° F (60° C) to 70° F (21° C) within two hours, and to 41° F (5° C) within four hours. Corrective actions are included in the HACCP Plan (Figure 4.7).

6- Establish a Record-Keeping System. The team will develop log sheets and records to be used by staff to verify that the HACCP system is being used. The log sheets should be simple to use

Figure 4.8 ■ HACCP Recipe for Stuffed Peppers

Menu Item: Stuffed Peppers   Yield: 24 portions

Ingredients	Amount	Potentially Hazardous Food
Green peppers, large	12	
Rice	10 oz	✔
Ground beef	2 1/2 lb	✔
Onion, fine dice	10 oz	
Chopped garlic	1 tsp	
Tomato sauce	3 pt	
Oil	3 oz	
Water		

Directions • CCP = Critical Control Point

1. Wash hands before beginning food preparation. Use clean and sanitized utensils and equipment.

2. Thaw ground beef under refrigeration (41°F/5°C or below), 2 days

3. Wash the peppers before cutting. Cut the peppers in half lengthwise. Remove seeds and core.

4. Blanch the peppers in boiling salted water or in a steamer for 4-5 minutes. (Peppers should be partially cooked but still crisp and firm.) Cool quickly in cold water.

5. Cook the rice by steaming. Cook to internal temperature of 165°F (74°C) at endpoint of cooking.

CCP 6. Cool to 41°F (5°C) within 4 hours. Use a cold water rinse to expedite cooling.

7. Sauté the onions and garlic in oil until lightly browned. Cool.

8. Combine rice, onions, meat and tomato sauce. Season to taste.

9. Fill each pepper half with 4 oz meat mixture.

10. Arrange peppers in baking pans and add about 1/4" water to the bottom of the pan (do not pour it on the peppers).

CCP 11. Bake at 350°F (175°C) for 30-35 minutes until internal temperature is 165°F (74°C) and the tops are browned.

CCP 12. Hold for service at 140° F (60° C) or above.

and to understand. All staff must be trained in the proper techniques for monitoring critical control points and recording data on the log sheets.

7. Establish Procedures to Verify that the System is Working. The team agrees to meet on a regular basis to review HACCP logs, to ensure that records are being completed and to identify any necessary changes.

Key Concepts

■ The HACCP system is a preventive food safety program that greatly reduces the likelihood of foodborne illness.

■ A hazard is unacceptable contamination, unacceptable growth, or unacceptable survival of harmful microorganisms.

■ Essentially, HACCP involves identifying where and how food could become unsafe (hazard analysis) and establishing standard operating procedures to protect food (critical control points).

■ Staff training is essential to the successful implementation of HACCP.

■ A critical control point (CCP) is an operation or a step of an operation, at or by which a preventive measure can be exercised that will eliminate, prevent or minimize a hazard that occurred prior to the that point.

■ Critical limits are specified limits or characteristics of a physical, chemical or biological nature.

■ Critical limits are observable and measurable.

■ When monitoring shows that a critical limit has not been met, there must be a corrective action to follow.

■ A record-keeping system documents HACCP activities and includes time and temperature monitoring records.

■ It is important to verify that a HACCP system is working over time.

Case Study

Lee is developing a HACCP plan for the China Moon Restaurant. Lee is working on a flow chart for one of the restaurant's signature items, Hunan Pork with Vegetables. The pork loin is received in a frozen form. The pork is thawed and then cut into cubes. Fresh vegetables are used in the recipe. The preparation method for the pork is stir frying. The Hunan Pork is prepared in small batches and held for 30 minutes to keep up with customer demand during peak serving times. For quality reasons, the leftovers are never kept. Using the flow chart on the next page, develop a flow chart for the Hunan Pork with Vegetables recipe.

☛ next page

Ingredients for Hunan Pork with Vegetables

Pork loin, cubed

Green peppers

Carrots

Onions

Soy sauce

Salt

Water chestnuts, canned

Bamboo shoots, canned

Ginger, fresh

Garlic, fresh

Oil

Cayenne pepper, dried

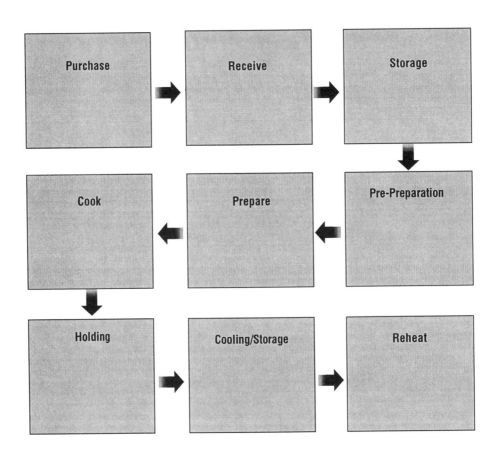

Review Questions

1. The first step in establishing a HACCP system is:

    A. determining critical limits

    B. identifying critical control points

    C. assessing hazards

    D. setting up a record-keeping system

2. A critical control point (CCP) is:

    A. a point at which loss of control may result in an unacceptable risk

    B. a point at which loss of control does not lead to an unacceptable risk

    C. failure to meet a required critical limit

    D. presence of a hazard or health risk

3. Which of the following is probably NOT a potentially hazardous food?

    A. egg custard

    B. French toast

    C. lettuce and tomato salad

    D. meatloaf

4. A technique used to identify critical control points (CCPs) in the flow of food is:

    A. menu analysis

    B. self-inspection

    C. flow charting

    D. quality control

5. Flow charts and HACCP recipes should be kept:

    A. in the manager's office

    B. in the food production area

    C. in the dry storage/receiving area

    D. in the training classroom

6. Checking to see that criteria, such as critical limits, are met in the HACCP system is called:

    A. documentation

    B. control

    C. hazard analysis

    D. monitoring

7. Confirmation that a HACCP system is working over time is called:

    A. verification

    B. consultation

    C. analysis

    D. documentation

8. The temperature of hot food being held is monitored every hour. The most recent reading is 120°F (49°C). What is an appropriate corrective action?

    A. dispose of the food immediately

    B. increase the temperature of the hot holding unit

    C. reheat to 155°F (68°C)

    D. reheat to 165°F (74°C)

9. What is the most appropriate critical limit or standard for cooling soup?

    A. cool to 45°F (7°C) within two hours

    B. cool from 70°F (21°C) to 41°F (5°C) within four hours

    C. cool in shallow pans

    D. cool by ice bath method

10. In which operational step would a critical control point (CCP) most likely be present?

    A. purchasing

    B. receiving

    C. storage

    D. preparation

Management and Personnel

Chapter Objectives

- Discuss the role of managers in protecting food

- Describe training methods and resources for foodservice personnel

- Identify hygiene standards for foodservice workers

- Describe proper hand washing techniques

- Identify how to prevent disease transmission by employees

Protecting food requires that foodservice operations are directed by trained and capable managers. And, it requires that employees understand and perform the tasks of purchasing, receiving, storing, preparing, and serving safe food. Personal hygiene — especially hand washing — and many other personnel policies are good examples. Never to be underestimated, employee effectiveness depends on training.

Foodservice Managers

The manager is responsible for making many daily decisions that directly impact the safety of food served. The manager decides where to purchase food and may be directly responsible for receiving orders. The manager must observe the practices of staff to ensure that safe food preparation practices are followed. The manager is responsible for hiring, orienting and training staff, and for ensuring that staff follow all policies and procedures to maintain a sanitary foodservice operation. The manager is responsible for implementing HACCP food safety systems and verifying the success of HACCP programs. To achieve these important objectives, managers must be trained in food protection practices. Many state and local health departments require that a manager certified in food protection practices be on the premises at all times.

Training and Developing Personnel. Foodservice managers have an obligation to train every member of the foodservice team in safe food preparation practices. Training improves the likelihood of compliance with food handling policies and practices, reduces the need for constant supervision, and improves employee self esteem and job satisfaction. Training is an important responsibility and trainers should be carefully selected based on job knowledge and ability to facilitate learning.

In this chapter, we will examine three major types of training:

1) New employee orientation

2) On-the-job or job instructional training and

3) Group training.

All are of importance to improving and maintaining the skills and knowledge of foodservice employees.

Orientation. The process of orientation prepares the new employees for their work responsibilities. Often an orientation includes the introduction to co-workers, review of important policies and procedures, and a review of job responsibilities. Orientation is an opportunity to introduce new employees to the operation's food protection practices.

An orientation should be carefully planned by assessing what a new employee needs to know. A survey of current employees and a review of the job description will help you to identify these critical areas. Planning includes determining the best method to cover the material and developing an orientation checklist to ensure a thorough and complete orientation process. Conducting the orientation may be a responsibility shared by the foodservice manager and foodservice staff. Evaluation is an important, but often overlooked, process for determining whether the orientation objectives for the new employee were met.

On-the-Job Training. On-the-job training or job instructional training is the training the new employee receives related to the performance of specific job duties. For instance, the on-the-job training program for a new dishwasher would focus on the proper steps in manual and automatic dish washing. The new employee would learn how to perform these job tasks by working with an experienced employee. The instruction would be accomplished through observation and return demonstration by the new employee, while the experienced employee provides coaching and feedback.

Group Training. In addition to on-the-job training, it is necessary to offer employees group training. Examples of group training are described in Figure 5.2.

Trainer Selection. It is very important to carefully select employees who will be training other staff members, and to prepare them for this job responsibility. Some employees have no experience in training, no time to do it, and no desire to participate. The qualities of effective on-the-job trainers include:

■ positive attitude

■ effective one-on-one communications skills

Figure 5.1 ■ Four Step On-the-Job Training Procedure

STEP 1 ■ Prepare

Arrange Work Area: Arrange the work area to allow the trainer to stand next to the trainee; collect any needed tools; remove distractions if possible.

Put Trainee at Ease: Reduce the trainee's anxiety, assure him/her that you are there to help and encourage the trainee to ask questions if he/she doesn't understand any directions.

STEP 2 ■ Present

Demonstrate the Job Function: Show the trainee how to do the job function by actually doing it and allowing him/her to observe. If the trainee desires, he/she may take notes during this phase.

Explain Key Points: While demonstrating, explain key points to the trainee. Attempt to answer the questions: Who, What, When, Where, Why, and How.

Repeat the Demonstration: The trainer again performs the task as the trainee observes. This reinforcement increases the likelihood that the trainee will retain the new information.

STEP 3 ■ Practice

Let the Trainee Attempt Parts of the Job: Allow the trainee to practice part of the job. After he/she attempts to do part of the job, provide constructive feedback, coaching, and suggestions.

Let the Trainee Perform the Entire Job Under Observation: The trainer now provides feedback periodically rather than frequently and encourages the trainee to begin acting independently.

STEP 4 ■ Follow Up

Conduct Periodic Progress Checks: The trainer steps back and only spot checks the trainee's progress.

Allow the Learner to Work Independently: At this point, the trainee has formally completed the training topic, but is encouraged to seek out the trainer if assistance is needed. Complete the training checklist.

■ interest in helping others learn and grow

■ job knowledge

■ patience

■ sense of humor

■ listening skills

Training Resources. The foodservice manager may require instructional materials to assist in providing food safety training to staff. A variety of resources are available from groups and organizations such as:

Figure 5.2 ■ Examples of Group Training Methods

Lectures • Presentation of information by a trainer or instructor. Requires little learner involvement. Effective for presenting new information. Disadvantage is that learner may not retain information.

Role Playing • "Acting out" of behaviors or practices. A participatory approach to training. Can be used to allow employees to practice new behaviors or habits during the training session.

Case Studies • Analysis of real life events or incidents. Incidents become a tool for learning and group discussion. Encourages the learner to apply knowledge to actual situations that may be encountered on the job.

Games and Simulations • Structured activities that encourage a high level of competition and participation. Excellent technique to review familiar material and reinforce knowledge.

- Professional associations in the foodservice and hospitality industry

- State departments of public health

- Federal agencies such as the Food and Drug Administration, United States Department of Agriculture and Centers for Disease Control and Prevention

- Cooperative Extension Program

- Local colleges, universities and technical schools

- Vendors and suppliers to the foodservice industry

Foodservice Personnel

Foodservice personnel must observe personal hygiene practices to reduce the likelihood of contaminating food and must also observe safe work practices. It is the responsibility of foodservice managers to develop and implement policies for foodservice worker hygiene. It is the obligation of every foodservice worker to observe standards of personal hygiene in order to ensure food safety. Foodservice workers should observe the following standards for personal hygiene:

- Bathe daily.

- Cover mouth when coughing or sneezing and then wash hands.

- Do not eat, drink or use tobacco or gum while preparing food. Wash hands after doing these activities on break.

- Restrain and cover hair.

- Frequently wash hands following established procedures.

Lavatory Facilities and Supplies. Foodservice personnel must be provided with a sink specifically intended for washing hands. It is not acceptable to wash hands in sinks intended for food preparation or to perform food preparation activities in hand sinks.

Figure 5.3 ■ Hand Sink with Infrared Sensor

Courtesy: Fisher Manufacturing Co.

At least one sink must be provided, and more may be required to ensure convenient use by employees. The sinks should be easily accessible by employees.

Hand washing sinks should be available in food preparation, foodservice, and warewashing areas, as well as in restrooms. A hand washing lavatory must provide water at a temperature of at least 110° F (43° C) through a mixing valve or faucet. Some models of hand sinks are equipped with infrared sensors or flow control devices, which eliminate the need for staff to touch sink faucets. The lavatory must be supplied with a hand cleanser and materials for hand drying such as: 1) individual, disposable towels, 2) a continuous towel system that supplies the user with a clean towel, or 3) a heated-air drying device. When food exposure is limited and hand washing lavatories are not conveniently available — such as in some mobile or temporary food establishments or at some vending machine locations — employees may use chemically treated towelettes for hand washing.

Hand Washing and Hand Care. Preparing and serving foods with hands is a common way to transfer pathogens and other food hazards from people to food. Hand washing is the most important step that all foodservice workers can take to protect food. All employees should be trained in proper hand washing procedures. The most effective way to determine employee compliance with hand washing standards is to observe on-the-job practices. Also, reminders posted in restrooms, break areas and at sinks can help to reinforce this important practice.

Hands must be kept clean before, during and after preparing foods. According to the FDA Model Food Code, hands should be washed:

Figure 5.4 ■ Use of a Nail Brush

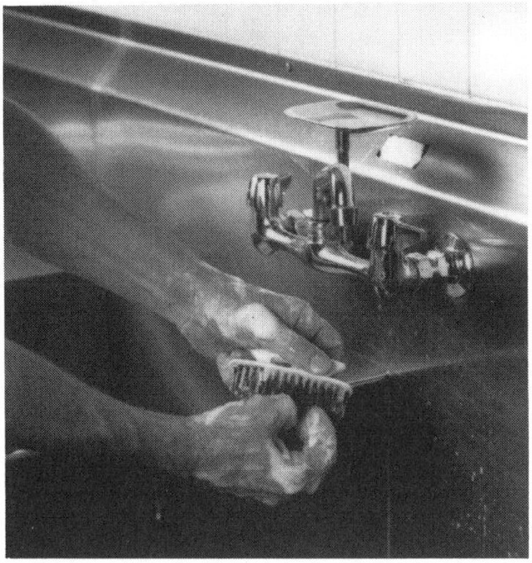

Courtesy: *Tucel Industries, Inc.*

Figure 5.5 ■ Sample Hand Washing Procedure

 Use warm, running water

 Wet hands & arms up to elbows

 Apply soap or detergent

 Rub hands and forearms vigorously for 20 seconds

 Scrub between fingers and under nails

 Rinse thoroughly under running water

 Dry hands and arms using a single service towel or hot air dryer

 Use the towel to turn off the faucet

- After touching bare human body parts

- After using the restroom

- After coughing, sneezing, using a tissue, using tobacco, eating, or drinking

- After handling soiled equipment or utensils

- Before food preparation

- During food preparation

- When switching between raw foods and ready-to-eat foods

- After engaging in any activity that might contaminate hands

Foodservice personnel must clean their hands, wrists, forearms and other exposed parts of their body with a cleaning compound (antibacterial soap recommended). Proper hand washing requires application of the cleaning compound and vigorously rubbing the hands for 20 seconds, followed by a rinse with potable water. Ensuring the cleanliness of areas underneath fingernails and in between fingers is particularly important. A nail brush should be used to clean under fingernails, particularly after handling raw food which may be trapped under the fingernails, or after using the restroom. Foodservice workers must keep their fingernails trimmed, filed and maintained so the edges and surfaces are cleanable and not rough.

Foodservice workers should limit jewelry to one single band ring. Jewelry can trap microorganisms and is difficult to keep clean. Skin lesions, cuts on the hands, wrists, or exposed portions of the arm must be covered with an impermeable cover such as a finger cot or stall. If on the hands or wrists, a disposable glove should be worn over the impermeable cover.

Disposable Glove Usage. The use of disposable gloves in foodservice operations is increasingly common. It is important to note that the wearing of gloves is not a substitute for appropriate, effective, thorough and frequent hand washing. Use single-use gloves that are stored and dispensed in a manner that prevents contamination. The glove should be intact and free of tears or other imperfections. Replace gloves at least hourly, when changing food preparation tasks, or after sneezing, coughing, touching hair, face or non-disinfected surfaces.

Clothing. Foodservice workers should always wear clean clothing to prevent contamination of food, equipment, utensils and linens. If employees are expected to change into a uniform after reporting to work, provide a suitable changing area separate from food preparation and storage areas. Employees should not bring personal items such as coats, hats, and purses into food preparation areas. To further minimize the risk of foodborne illness transmission from employees or employee work habits, protective clothing should be worn to keep body parts from coming in contact with exposed foods.

Hair restraints and beard restraints should be worn to reduce contact with human hair. Aprons and disposable gloves may also be used to reduce the transfer of microorganisms to an exposed food. Aprons and disposable gloves should be changed frequently.

Foodservice Practices

Avoid Hand Contact with Food. In addition to hand washing, the risk of foodborne illness can be reduced by eliminating direct contact of hands with ready-to-eat food. Except when washing fruits and vegetables, foodservice employees may not contact exposed, ready-to-eat food with their bare hands and should use suitable utensils such as deli tissue, spatulas, tongs, single-use gloves or dispensing equipment. Foodservice personnel should also minimize bare hand and arm contact with exposed food that is not in a ready-to-eat form.

Tasting of Food. Foodservice personnel should be trained in proper tasting techniques. To prevent contamination, food should not be tasted while standing over food. A foodservice worker may not use a utensil more than once to taste food that is to be sold or served.

Handling of Utensils. Employees should avoid touching the eating ends of the utensils, the tops or insides of drinking glasses, the surfaces of plates or any other utensil or service ware in a manner in which the hands might contaminate the item.

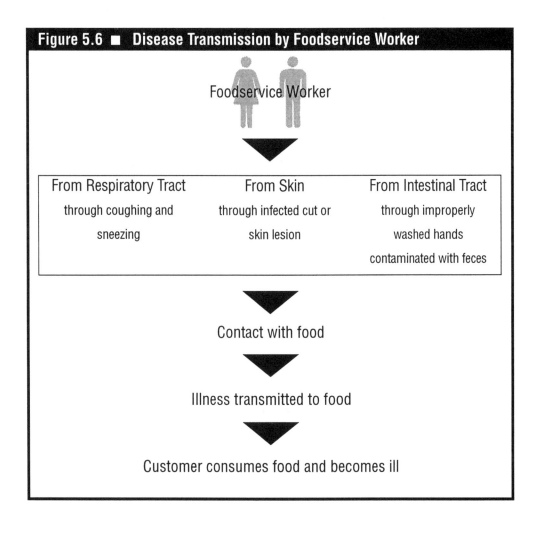

Figure 5.6 ■ Disease Transmission by Foodservice Worker

Foodservice Worker

From Respiratory Tract	From Skin	From Intestinal Tract
through coughing and sneezing	through infected cut or skin lesion	through improperly washed hands contaminated with feces

Contact with food

Illness transmitted to food

Customer consumes food and becomes ill

Figure 5.7 ■ Restrictions for Preventing Disease Transmission

Illness	Restriction	Comments
Abscess, skin lesions, boils (infected)	No direct customer contact or food preparation; use impermeable cover	May return to work when drainage ceases and lesion has healed or employee has a negative culture; can work if affected area is covered by impermeable cover.
Diarrhea (acute stage)	No direct customer contact or food preparation	May return to work when symptoms resolve and infection with Salmonella*, Shigella* or Campylobacter** is ruled out
Hepatitis A	No direct customer contact or food preparation	Until 7 days after onset of jaundice (local health department should be contacted)
HIV/AIDS	No restrictions	Counsel employee to ensure he/she observes safe work and hygiene practices
Lacerations, abrasions and burns (non-infected)	May have customer contact and prepare food with proper precautions	Proper precautions include keeping affected area covered by clean bandage and wearing disposable gloves if on hands or lower arms
Respiratory Infection (no fever)	May have customer contact and prepare food with proper precautions	If no fever, wash hands frequently and observe safe work practices. If excessive coughing or fever, restrict from foodservice operation.
Strep Throat	No direct customer contact or food preparation	Until 24 hours after effective treatment is started

*Restrict until 2 consecutive negative cultures, at least 24 hrs apart.
**Restrict until symptoms resolve or appropriate antibiotic treatment has been in effect for 48 hrs.

Preventing Disease Transmission by Employees. Foodservice personnel should never be permitted to work when they possess symptoms of diseases that can be transmitted through food. Although not required by the FDA Model Food Code, health examinations for foodservice workers may be required by local health codes. During employment, foodservice employees should report any illness to the foodservice manager. Employees with obvious symptoms such as sneezing, coughing, open skin lesions, fever, or gastrointestinal illness should never prepare food or work in any area of the foodservice operation. Any employee suffering from a foodborne illness should not be permitted to work until advised by a physician. Dangerous pathogens can be transmitted even if the employee is not working in food preparation.

Certain foodborne illnesses are primarily the result of infected foodservice workers or carriers contaminating food. The bacteria Staphyloccocus aureus, for example, is found on the skin, in infected cuts and pimples, in our throats and noses. The bacteria can be transmitted from the foodservice worker to food, where it can multiply to numbers sufficient to produce a toxin. Once served, the infected food causes the client to become ill. Viruses cannot multiply in food but can be transmitted through food. Examples of such viruses include Hepatitis A and the Norwalk virus. Other viruses, such as the HIV virus, are very fragile outside of the human body and cannot be transmitted by food. Foodservice workers with HIV or AIDS should not be excluded unless they are suffering from another illness or infectious disease.

Key Concepts

■ The manager is responsible for making many daily decisions that directly impact the safety of food served.

■ The manager must observe the practices of staff to ensure that safe food preparation practices are observed.

■ Training improves the likelihood of compliance with food preparation policies and practices, reduces the need for constant supervision, and improves employee self esteem and job satisfaction.

■ It is the obligation of every foodservice worker to observe standards of personal hygiene in order to ensure food safety.

■ Hand washing is the most important step that all foodservice workers can take to protect food.

■ A hand washing lavatory must be equipped to provide water at a temperature of at least 110° F (43° C) through a mixing valve or faucet.

■ Wearing of gloves is not a substitute for appropriate, effective, thorough and frequent hand washing.

■ Foodservice workers should avoid direct hand contact with ready-to-eat food.

■ Foodservice workers should always wear clean clothing and change aprons when soiled.

■ Foodservice personnel should never be permitted to work when they possess symptoms of diseases that can be transmitted through food.

Case Study

Charlotte is the foodservice manager for Ridgeview High School. One of the foodservice workers on staff, Robin, reported to work with flu-like symptoms including cough, sneezing, and a complaint of diarrhea. Robin has a poor attendance record and reported to work fearing that she would be disciplined for missing another day of work. Another staff member, Janet, cut her hand the previous day while using a chef's knife. She had to have several stitches in the laceration and has reported to work wearing a bandage on her hand. Charlotte is already short staffed and does not want to send any employees home. What should she do?

☞ 1. Should either employee be sent home?

☞ 2. What precautions could be taken to prevent disease transmission?

☞ 3. What are the possible risks if both employees are allowed to work?

Review Questions

1. The most important personal hygiene practice for a foodservice worker to follow would be to:

    A. wash hands

    B. restrain hair

    C. wear clean clothes

    D. eliminate gum chewing

2. A foodservice worker diagnosed with illness due to Salmonella typhi should:

    A. be restricted from working

    B. be permitted to serve food only

    C. be permitted to work in the dishroom

    D. continue working but wash hands frequently

3. What is the best way to promote proper food safety among employees?

    A. post reminder notices

    B. educate employees during orientation

    C. ask employees to remind one another

    D. provide ongoing training sessions and monitoring

4. After handling dirty dishes, foodservice workers should vigorously scrub hands for how many seconds?

    A. 5

    B. 10

    C. 15

    D. 20

5. When washing hands, foodservice workers should pay particular attention to:

    A. palms of hands

    B. under fingernails

    C. upper arms

    D. wrists

6. A person whose body harbors a disease-causing microorganism is called a(n):

    A. infector

    B. carrier

    C. inoculant

    D. toxigen

7. If disposable gloves are used:

    A. they will eliminate the risk of foodborne illness

    B. they should be changed once during the shift

    C. they can substitute for hand washing

    D. they should be changed as frequently as you wash your hands

8. The temperature of lavatory water for hand washing should be at least

    A. 90°F (32°C)

    B. 100°F (38°C)

    C. 110°F (43°C)

    D. 120°F (54°C)

9. A group training method that is appropriate for presenting new information, but allows only limited participation for learners is:

    A. case study

    B. lecture

    C. role playing

    D. simulation

10. Which of the following can NOT be transmitted to food by foodservice personnel?

    A. Staphyloccocus aureus

    B. Hepatitis A

    C. Norwalk virus

    D. HIV

Cleaning and Sanitizing

Chapter Objectives

■ Discuss the importance of maintaining a clean and sanitary environment

■ Identify how to develop a cleaning program

■ Distinguish between clean and sanitary

■ Identify cleaning products and effective cleaning techniques

■ Identify sanitizing products and effective sanitization techniques

■ Review the principles of Integrated Pest Management

■ Discuss waste management and disposal

Maintaining a clean environment for food preparation is an important part of food safety. The presence of bacte- ria and the likelihood of foodborne ill- ness increase if a foodservice operation fails to use appropriate cleaning and sanitizing techniques. Equipment, food contact surfaces, and utensils must be clean to sight and touch. The food contact surfaces of equipment should be free of any encrusted grease deposits and other soil accumulations. Non-food contact surfaces should be kept free of dust, dirt, food residue and debris. A philosophy of clean-as-you-go should be instilled in all foodservice staff. A cleaning program will most likely be effective if the foodservice manager incorporates a system of self-inspection and training.

Developing a Cleaning Program

Maintaining a clean foodservice operation requires that foodservice managers utilize their skills in planning, organizing, staffing, training, and directing. The manager must first determine *what* the cleaning needs of the operation are, and *when* these functions need to be carried out. The manager must then determine *who* will be responsible for carrying out the cleaning. Some cleaning may be done by contract personnel, but most will be completed by foodservice employees. The manager must provide adequate staffing so that employees will be able to carry out the planned cleaning activities. The manager must train employees in proper cleaning techniques and safe use of chemicals. Finally, the manager must evaluate the cleaning program on an ongoing basis to ensure its effectiveness and to correct any identified problems. A written checklist, such as that in Figure 6.1, can be developed to outline the cleaning program and employee responsibilities. A self-inspection checklist (Figure 6.2) can be used to evaluate the effectiveness of the cleaning program. Detailed instructions on proper cleaning techniques should be incorporated into employee training programs. An example of this appears in Figure 6.3.

Cleaning Products. Detergents are powerful dirt-removing cleaning agents. Types of detergents are described in Figure 6.4. A cleaning supply vendor can assist you in selecting the right chemical for the right job. Factors in selecting a detergent include:

- **dissolvability**: how well the detergent disperses with water

- **wetting ability**: how well it saturates the surface of the object to be cleaned

- **emulsification ability**: how well the detergent suspends the fats in water and keeps them suspended

- **free rinsing ability**: how well the detergent keeps soil from returning to the cleaned object

- **deflocculation ability**: how well the detergent puts lumps of food-soil in suspension in water

- **nontoxicity**: the extent to which the detergent is free of toxins and other elements harmful to humans

- **corrosiveness**: how much of a corroding effect the detergent has on the surface of the item to be cleaned

Wiping Cloths. Cloths that are used for wiping food spills should be used for no other purpose. Cloths used for wiping food should be stored in a sanitizing solution between uses — unless a dry cloth is used. Dry or moist cloths that are used with raw animal foods must be kept separate from cloths used for other purposes, and moist cloths used with raw animal foods must be kept in a separate sanitizing solution.

Clean vs. Sanitary. Clean does not always mean sanitary. **Cleaning** is the removal of dirt or debris by physical and/or chemical means. **Sanitary** means free of harmful levels of micro-organisms. It almost impossible to sanitize equipment and facilities unless they are first clean. Clean food-service equipment may appear safe to use, but if it is not sanitized, then a hidden threat to food safety exists. Equipment and utensils must be sanitized before use and after cleaning.

Sanitizing. Sanitizing is the destruction of microorganisms that remain on equipment and work surfaces after

Figure 6.1 ■ Sample Daily Cleaning Schedule

Daily Cleaning Schedule

Area:_____ Week of:_____

Item	Person Responsible	Initial When Completed						
		S	M	Tu	W	Th	F	S
Grill	Cafe Worker 1							
Fryer	Cafe Worker 1							
Reach-in Refrigerator	Cafe Worker 1							
Food Warmer	Cafe Worker 2							
Steam Table	Cafe Worker 2							
Salad Bar	Cafe Worker 3							
Serving Line	Cafe Worker 3							
Soda Dispenser	Cashier							
Dining Room Table	Cashier							
Floor	Utility Aide							
Garbage Cans	Utility Aide							

Figure 6.2 ■ Sample Cleaning Self-Inspection Checklist

Area	Clean ✔		Corrective Actions
	Yes	No	
Kitchen Floors			
Walls			
Ceilings			
Baseboards			
Hoods and Vents			
Light Fixtures			
Counters			
Filters			
Fans			
Dining Room Floors			
Walls			
Ceilings			
Baseboards			
Light Fixtures			
Dry Storage Area Floors			
Walls			
Ceiling			
Baseboards			
Light Fixtures			

washing and rinsing. There are two major methods of sanitizing: 1) heat sanitization and 2) chemical sanitization.

■ **Heat Sanitization:** Exposing an object to sufficiently high heat for a sufficient period of time will sanitize it. According to the FDA Model Food Code, cleaned food contact surfaces can be sanitized by immersion in water that is 171°F (77°C) or above for at least 30 seconds.

■ **Chemical Sanitization:** *Sanitizers* are chemicals that destroy harmful pathogens. The dilution or strength of a sanitizing solution is measured in parts per million (ppm). Dispensing equipment for sanitizers should be calibrated on a regular basis and the strength of a sanitizing solution should be checked with a chemical test strip on a daily basis. Chemical test strips are often available from the chemical supply vendor.

Manual Cleaning and Sanitizing. Manual cleaning and sanitizing is generally done with a three-compartment sink. Some municipalities are now requiring a four-compartment sink. The additional compartment is designated for pre-scraping. If you are using a three-compartment sink, remember to pre-scrape utensils and cooking equipment before placing it in the first sink compartment. The steps in manual cleaning and sanitizing using a three-compartment sink are:

1. **Pre-clean** to remove loose food soil.

2. **Wash** in the first tank using an approved detergent with a water temperature of 110°F - 120°F (43°C - 49°C).

3. **Rinse** in the second tank, using clear water.

4. **Sanitize** in the third sink using an approved sanitizer at the appropriate concentration and temperature.

Figure 6.3 ■ Sample Cleaning Procedure for Floors

How to Clean Floors

Equipment Needed:
broom
dust pan
2 mops
2 mop buckets

Supplies Needed:
heavy-duty detergent

WHAT TO DO	HOW TO DO IT
1. Sweep	• Sweep the floor with a broom, making sure to get under equipment. • Minimize dust by using short, controlled strokes. • Use dust pan to pick up debris.
2. Set up mop buckets	• Fill each bucket 3/4 full with hot tap water. • Put the appropriate amount of cleanser in one of the buckets.
3. Apply cleanser to floor	• Dip first mop into cleanser solution and wring out. • Mop small section of floor using a figure-eight motion. • Allow detergent to loosen soil for a few minutes.
4. Rinse floor	• Dip the second mop into the rinse water, wring out, and use the mop to pick up the dirt and cleaning solution. • Change detergent solutions and rinse water often.

SAFETY CONSIDERATIONS

Check product label and refer to MSDS for precautions and emergency procedures in the event of an accident.

Follow manufacturer's directions on detergent label to ensure the proper dilution.

Use "wet floor" signs to prevent accidental slips and falls.

5. Allow dishes to air dry and store in a clean, dry, protected area.

Remember to change the water as necessary and to clean the sink after each use. Use cleaned and sanitized drain boards to stack the equipment and utensils during air drying.

Mechanical Cleaning and Sanitizing.
Mechanical cleaning and sanitizing is accomplished with the use of a dishmachine. While there are a variety of different sizes and types of dishmachines, the following steps will apply to most operations. These six steps are necessary in any mechanical

---

Figure 6.4 ■ Types of Detergents

Alkaline Detergents ■ The most commonly used detergents. General-purpose alkalines are only mildly alkaline and are usually applied to walls, ceilings, and floors. Heavy-duty types are highly alkaline for use in dishwashing machines.

Abrasive Cleaners ■ Effective in scouring off rust, grease and heavy soil but can scratch certain surfaces such as stainless steel. Use caution with abrasive cleaners, as scratches in food contact surfaces can harbor bacterial growth.

Acid Cleaners ■ Acid cleaners have more specialized uses, such as deliming dish machines and removing water spots. Use acid cleaners with caution to avoid skin irritation.

Degreasers ■ Degreasers are highly alkaline and can remove grease build-up from floors, ovens, and vents. Degreasers are also known as solvent cleaners.

Figure 6.5 ■ Chemical Sanitizers

Product	Dilution in parts per million (ppm)	Water Temperature	Immersion Time ▼
Iodine	12.5 - 25	75° - 120° F (24 - 49° C)	30 seconds
Chlorine	50 - 100	75° F (24° C)	30 seconds
Quaternary ammonium compounds	180 - 200	75° F (24° C)	30 seconds

▼ general guidelines only; follow manufacturer's directions or state code

warewashing operation. Each step is equally important in obtaining clean and sanitary dishware.

1. **Separating:** Separate any items that will require special attention, such as items that are heavily soiled or have burned-on residue.

2. **Prescraping or preflushing:** Excess food soil and pieces of paper straws or napkins will clog the dishmachine, scrap trays, pump and wash arms — and reduce their effectiveness. Excess food soil in the wash tank also uses up more detergent and requires more frequent changes of

water in the wash tank.

3. **Racking:** Proper racking of dishes is essential for getting good results. All of the racks should be filled with similar items. A rack should contain cups only, or plates only, or bowls only, etc. Avoid mixing loads. The dishmachine works by spraying water at the dishes, and if the water cannot reach all of the dish surfaces, the dishes can't get clean.

4. **Washing**: Washing dishes requires a properly operating dishmachine and the proper detergent usage. You must be sure to use the right detergent, make sure that the machine is set up properly, and see that it is free of paper, straws, and excess food.

5. **Sanitizing Rinse**: Rinsing is done automatically and requires only the proper water flow, proper water temperature, and the proper rinse aid. Visually inspect the machine to ensure that the rinse jets are not clogged. Verify that rinse temperature is reaching the appropriate point and that the rinse injector is supplied with the proper rinse aid.

6. **Air Drying**: Air drying the dishes requires nothing but time. Wait until the dishes are completely dry before putting them in a clean, dry storage area. Avoid stacking utensils within one another until they are completely dry.

Dish and Equipment Storage. After cleaning and sanitizing, all dishes and utensils must be stored dry and in clean, dust-free areas above the floor and protected from dust, splashes, spills and other forms of contamination.

Chemical Storage and Container Use. Cleaning and sanitizing chemicals must be stored away from food preparation and storage areas. Chemicals should be stored in their original containers and be clearly labeled. Containers used to hold and dispense chemicals during use should also be clearly marked as cleaner or sanitizer containers, and should never be re-used for food storage.

Common Pests and Pest Control

A safe food operation is free of insects, birds, and other sources of contamination. Measures must be taken to prevent contamination of food by pests. Insects and rodents are disease-carrying pests that can render food unfit for human consumption.

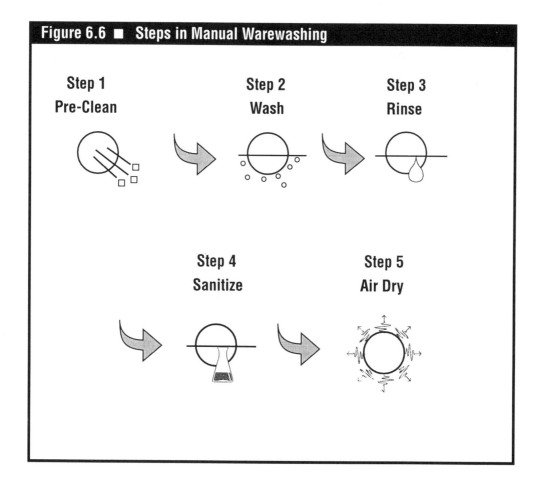

Figure 6.6 ■ Steps in Manual Warewashing

Step 1
Pre-Clean

Step 2
Wash

Step 3
Rinse

Step 4
Sanitize

Step 5
Air Dry

Figure 6.7 ■ Mechanical Dishmachine Temperatures

Type of Machine	Temperature of Wash Solution	Temperature of Sanitizing Rinse
Stationary rack single temperature machine	165° F (74° C)	165° F (74° C)
Stationary rack dual temperature machine	150° F (66° C)	180° F (82° C)
Single tank conveyor dual temperature machine	160° F (71° C)	180° F (82° C)
Multi-tank conveyor multi-temperature machine	150° F (66° C)	180° F (82° C)

Note: The actual temperature attained within a hot water sanitizing machine (at utensil surface) should be at least 160° F when measured by a maximum registering thermometer or temperature test strip.

Cockroaches. Cockroaches are a common pest in foodservice operations, and they represent a significant threat. Cockroaches are carriers of disease and can multiply rapidly. Cockroaches are most active at night and in dark areas, where there is less disturbance from people. While there are many different species of cockroaches, three types are frequently found in foodservice operations. German cockroaches are pale brown with two dark-brown stripes behind the head. They seem to prefer warmer places in the foodservice operation. American cockroaches are the largest species in the United States and are reddish brown. They prefer open, wet areas and spaces that are slightly cooler than the favorite spots of German cockroaches. Oriental cockroaches are shiny and dark brown to black. They prefer conditions similar to those of the American cockroach.

Housefly. The common housefly is the greatest threat to food safety. Flies feed on animal and human wastes and then transmit pathogenic organisms as they travel from item to item. Houseflies are numerous during late summer and early fall, since the population has been growing during the warmer weather.

Rats and Mice. Rats and mice can force their entry into a building through small openings. Both rats and mice are skilled swimmers and can enter a building through floor drains and toilet bowl traps. Rats and mice are both nocturnal creatures, being most active at night. One indication

of an infestation is the presence of fecal droppings in the foodservice operation. Another sign is the presence of gnaw marks on food and food containers in storage.

Birds. Birds such as pigeons, sparrows, and starlings are another pest of concern to the foodservice manager. Birds can enter a building through doors and windows that do not have screens and through ventilation openings. Birds can be carriers of mites and harmful microorganisms such as Salmonella.

Pest Control. Pest control is the reduction or eradication of pests. One of the safest and most effective methods of fly control is the use of insect light traps. Light traps must be enclosed so that pests entering the trap are contained within and cannot fall out into the foodservice operation. Mechanical traps can be used with birds and rodents. Glue traps are also effective in capturing mice.

Pesticides, insecticides and repellents are other control techniques. Extreme caution should be exercised when using chemical control methods to prevent contamination of food. Only chemicals approved for application in foodservice operations may be used. Further, only staff with the proper training and knowledge should be involved in the application of chemical control methods. Check local regulations for further restrictions on the use of pest control devices.

Pest control cannot be accomplished effectively unless and until proper cleaning has occurred. One method of pest control by itself is not often effective. For this reason, many foodservice operations use an approach called *integrated pest management (IPM)*. IPM uses a variety of techniques and focuses on prevention. Foodservice operators should implement an IPM including these elements:

■ Restrict access into the facility with screens and building maintenance.

■ Reduce any food source for pests through good cleaning and sanitization practices.

■ Control pests by working with a licensed pest control operator (PCO).

All openings outside of the building must be closed to restrict access to the rodents. Screens in doors and windows can help to restrict the access of insects. Self-closing doors and proper seals around doors also reduced the accessibility of the foodservice operation to pests. All food storage and preparation areas must be kept in a clean and sanitary condition. Food spills are a source of food for pests. Food stored against the wall creates places for pests to hide, nest and multiply. Garbage cans in the facility and dumpsters outside should be kept closed or covered at all times and need to be cleaned frequently.

Waste Management. Proper management of waste and garbage will help to maintain a clean, sanitary foodservice operation and to prevent pest infestations. Garbage cans should be sturdy and easy to clean. Garbage cans should be covered with tight-fitting lids. Garbage cans should be cleaned on a routine basis, and garbage cans with cracks or leaks should be replaced immediately.

Dumpsters should also have doors that are tight-fitting. All garbage should be in plastic bags and not loose in the dumpster. The dumpster area should be cleaned on a regular basis, and garbage should be picked up at regular intervals to prevent excess odors and reduce food sources for pests.

Key Concepts

■ Maintaining a clean environment for food preparation is an important part of protecting food.

■ A philosophy of clean-as-you-go should be instilled in all foodservice staff.

■ A cleaning program will most likely be effective if the foodservice manager incorporates a system of self-inspection and training.

■ Detailed instructions on proper cleaning techniques should be incorporated into employee training programs.

■ Cleaning is the removal of dirt or debris by physical and/or chemical means.

■ Detergents are powerful dirt-removing cleaning agents.

■ Sanitary means free of harmful levels of microorganisms.

■ There are two major methods of sanitization — heat sanitization and chemical sanitization.

■ Sanitizers are chemicals that destroy harmful pathogens.

■ Integrated pest management (IPM) uses a variety of techniques to control pests and focuses on prevention.

Case Study

Doris is the foodservice director at Green Gardens Health Center, a long term care facility. Doris observes an employee preparing for the noon meal. The employee pulls a long pan of raw chicken that was previously breaded from the refrigerator. The employee places the individual pieces of chicken into a wire basket and into the broaster. After several minutes, the employee carefully checks the temperature at the endpoint of cooking. The employee then places the chicken back into the long pans and places it on the steam table to hold. Another employee uses the blender to puree beef stew. Doris watches as the employee takes the blender to the pot and pan sink and cleans it using the detergent and rinse water only. The employee then places the blender back on the base and continues working.

☞ 1. What mistake did the first employee make?

☞ 2. What mistake did the second employee make?

☞ 3. What risks to food safety were created as a result?

☞ 4. What would have been a safer practice in each case?

Review Questions

1. The first step in manual cleaning and sanitizing is to:

    A. rinse

    B. air dry

    C. scrape/sort

    D. sanitize

2. Degreasers are also known as:

    A. abrasive cleaners

    B. detergents

    C. solvent cleaners

    D. acid cleaners

3. A chemical intended to delime dishmachines is:

    A. a detergent

    B. a solvent cleaner

    C. an acid cleaner

    D. an abrasive cleaner

4. According to the FDA Food Code, the minimum immersion time for most sanitizers is:

    A. 30 seconds

    B. 60 seconds

    C. 90 seconds

    D. 120 seconds

5. Sanitary means:

     A. free of harmful levels of microorganisms

     B. absence of microorganisms

     C. immersion in chemical solution

     D. free of debris or soil

6. Which of the following is NOT an appropriate sanitizer?

     A. alkaline

     B. chlorine

     C. quaternary ammonium

     D. iodine

7. The best way to prevent pest infestations in storage areas is to:

     A. apply residual insecticides to all unrefrigerated food packages

     B. install lighted electronic "zappers" on ceilings of selected storage areas

     C. store dry goods in cool, dry areas away from the wall

     D. place glue traps in storage areas to deter insects from food supplies

8. The correct concentration of a sanitizer can be determined by:

     A. observing the color

     B. using a chemical test strip

     C. mixing with equal parts water

     D. using a thermometer

9. The temperature of wash water for manual cleaning should be:

    A. 75° - 100°F (24° - 38°C)

    B. 100° - 110°F (38° - 43°C)

    C. 110° - 120°F (43° - 49°C)

    D. 120° - 130°F (49° - 54°C)

10. The final step following mechanical cleaning and sanitizing is to:

    A. rinse

    B. sanitize

    C. air dry

    D. wash

Physical Facilities and Equipment

Chapter Objectives

- Describe considerations for floors, walls and ceilings

- Discuss the properties of heating, ventilation and air conditioning systems

- Identify food protection concerns with plumbing and waste water systems

- Describe food safety considerations for storage and preparation areas

- Identify standards for selection of safe foodservice equipment

- Discuss the installation and maintenance of foodservice equipment

---

The equipment used to prepare food and the physical facilities of the foodservice operation have a tremendous impact on the safety of food served. Cleaning and maintenance are important considerations when designing a foodservice facility and when purchasing equipment. A well designed foodservice operation utilizes materials that are durable, non-porous and easy to clean. Regulations for foodservice operations have changed considerably over the last several years, so it is important to consult new standards before making any construction, remodeling or major purchasing decision. The sanitary design of a foodservice facility is often governed by a variety of laws, including public health, building and zoning. The primary consideration for the sanitary design of foodservice facilities is cleanability. *Cleanability* is the extent to which an item is accessible for cleaning and inspection and ease in which soil can be removed by normal cleaning methods. Some examples of how food safety can be part of a facility design are:

■ Using building finishes that are durable and easy to clean

■ Equipment that is attached to the wall, eliminating legs, which

makes it easier to clean under the equipment

■ Equipment racks with a minimum number of legs

■ Garbage disposals in work areas to facilitate waste disposal

■ Portable shelf designed under tables that can be cleaned easily

Construction Materials and Considerations

Floors, Walls and Ceilings. The materials chosen for floors, ceilings and walls are selected on the basis of ease in cleaning, resistance and durability. Utility service lines should not be unnecessarily exposed and should not obstruct or prevent the cleaning of floors, walls, and ceilings. Walls and ceilings should be light in color to distribute light and to make any soil more noticeable.

Floors. The floor in food preparation and utility areas should be easy to maintain, wear resistant, slip resistant, and non-porous. Common materials for floors are quarry tile, terrazzo, and sealed concrete. Other materials may be used in dining rooms. For instance, tightly woven carpeting would be acceptable in a dining room but not in a food preparation or storage area. Vinyl tile is not recommended, even for dining rooms, because it is difficult to maintain and wears out quickly. Ceramic tile is a common floor finish for public restrooms and other high traffic areas. Floor mats, if used, should be easy to remove and to clean.

Floor Drains. Floors that are flushed with water for cleaning or are subject to frequent liquid spills should have adequate drains to allow the liquids to drain off by themselves.

Walls. The best wall finish in food preparation areas is structural glazed tile or ceramic tile. Both withstand heat, grease, and frequent cleaning. Structural glazed tile is also resistant to impact from movable carts and equipment.

Floor and Wall Junctures. Coving is required at the juncture of floors and walls. Coving at a floor-wall joint facilitates cleaning by preventing accumulation of bits of food that attract insects and rodents.

Ceilings. A wide variety of construction materials is available for ceilings, including: acoustical tile, painted drywall, painted plaster; and exposed concrete. Acoustical tile is a common choice because it is economical and has sound-absorbing qualities. Special non-absorbent acoustical tiles have been developed.

Heating, Ventilation, Air Conditioning System Vents. Heating, Ventilation, and Air Conditioning (HVAC) systems should help maintain appropriate temperatures and air flow. The elements of an HVAC system — including hoods, vents and filters — should be inspected and cleaned on a regular basis to ensure efficient operation and to prevent the risk of a fire from grease build-up. All preparation and work areas where odors, fumes, or vapors accumulate should be vented to the outside. Heating, ventilating, and air conditioning systems should be designed and installed so that make-up air intake and exhaust vents do not cause contamination of food, food contact surfaces, equipment or utensils. Vents and hoods should be maintained in a clean condition and air filters, if used, should be changed regularly.

Plumbing. The plumbing system performs two very important functions. It brings potable water to the operation and it removes waste materials to a sewer or disposal plant. A major consideration for food protection is to avoid any connection between the two functions.

Fixtures and equipment used for food preparation, cleaning or sanitizing may not be directly connected to a sewer connection. This type of connection is called a **cross-connection**. A cross-connection can result in backflow and back-siphonage.

Backflow is a reserved flow of unsafe water into sinks and equipment, creating the likelihood of contamination. **Back-siphonage** occurs after a loss in pressure in the water supply and water is siphoned back into the drinkable water supply. An **air gap** is an unobstructed air space between an outlet of drinkable water and the flood rim of a fixture or equipment. If an air gap is not present, then the equipment should be equipped with a mechanical device to prevent backflow and back-siphonage. A foodservice operation should have at least one service sink with a floor drain for the cleaning of mops and for the disposal of similar liquid waste. All sewage must flow into a public sewer system or into a disposal system that meets local regulations and requirements.

Lighting. Lighting should be: 1) adequate so that dirt and soil are visible; 2) easy to clean and 3) bright enough to prevent accidents from poor lighting. Lighting is measured in foot-candles. Working areas require 50 foot-candles of light. In food preparation and warewashing areas, lighting must be shielded and shatter-resistant.

Special Area Considerations

■ **Dry Storage Areas**: Storage areas should be well ventilated, dry, and constructed of easy-to-clean surfaces. Concrete or tile floors, cement block walls with epoxy paint, and acoustic ceilings are common for all storage areas except those that are refrigerated. Three- or four-level metal shelving is commonly used in storage areas. Slotted or louvered shelves are recommended for storage areas because they permit proper air flow.

■ **Warewashing Areas**: Warewashing areas must be designed so that they are easy to sanitize and can withstand wet conditions. Common construction materials include slip-resistant quarry tile floors, ceramic or structural glazed tile

walls, and moisture-resistant acoustic ceilings.

■ **Preparation Areas**: A major consideration in preparation areas is work surfaces. A common material used in the construction of work surfaces is stainless steel. Stainless steel is non-corrosive, non-absorbent, and non-toxic — making it a very suitable material for use in foodservice operations. All food contact surfaces must be accessible for cleaning and sanitizing.

■ **Toilet and Hand Washing Facilities**: Toilets should be conveniently located but separate from food preparation areas. The number of toilets and hand washing lavatories required will vary by the size of the foodservice operation and is usually determined by state and local regulations. Separate restrooms for employees and customers are recommended. The doors to toilet areas must be tight-fitting and self-closing. Requirements for hand washing lavatories and supplies are discussed in Chapter 5.

Foodservice Equipment

Equipment Standards. Many utensils and products intended for household use are not appropriate for use in foodservice operations. A foodservice manager should purchase only equipment that was intended for use in the foodservice industry. In general, foodservice equipment should be designed to be easily cleaned, maintained, and serviced — either in the assembled and disassembled state.

One way to identify properly designed and constructed equipment is to look for the NSF seal. The NSF seal of approval is a recognized standard of acceptance for many pieces of equipment. This seal assures the foodservice manager that the equipment meets certain construction standards for sanitation and safety. Another important standard is the UL mark. The Underwriters' Laboratories (UL) indicates compliance of the equipment with electrical safety standards.

Figure 7.1 ■ NSF International Symbol

The following are examples of characteristics to look for when purchasing foodservice equipment:

■ Easy disassembly to facilitate cleaning

■ Durable, corrosion resistant and non-absorbent

■ Materials that do not impart any significant color, odor or taste to food

■ Smooth surfaces free of pits, crevices, ledges, bolts and rivet heads

■ Coating materials that are resistant to cracking and chipping

■ Rounded edges and internal corners with finished smooth surfaces

Refrigeration Equipment. Refrigeration equipment is important to food protection. Many outbreaks of foodborne illness can be traced to inadequate refrigeration. Previous standards placed the upper temperature limit for refrigeration at 45° F. The 1993 Food Code revised the standard to 41° F (5° C). The Food Code included a five-year period for foodservice operations with older equipment to comply. To ensure maximum performance of refrigeration units, follow these steps:

■ Check to see that gaskets and hinges fit tightly and that there is no air leaking.

■ Clean or replace the air filter.

■ Clean the condenser.

Some foodservice operations connect thermostats on refrigerators to security or other monitoring equipment. Maintenance or management personnel can be notified immediately if the equipment fails — rather than discovering food spoilage several hours after the fact. Thermal barriers made of flexible, overlapping, PVC strips can be mounted on the doors of walk-in refrigerators and freezers to help maintain temperatures even when the door is opened.

Figure 7.2 ■ Thermal Barrier for Refrigerator and Freezer Doors

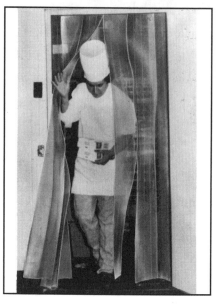

Courtesy: Curtron Industries, Inc.

Equipment Installation. Equipment that is fixed because it is not easily mobile should be installed so that it is spaced to allow access for cleaning along the sides, behind and above the equipment. Equipment that is not easily movable should be sealed to the floor or elevated on legs that provide at least 6" (15 cm) clearance between the floor and the equipment. Equipment mounted on concrete bases or on small steel legs can interfere with proper cleaning. Easier access for cleaning can also be provided by mounting equipment on wheels so it can be moved, or by using wall-mounted equipment.

Cutting Boards. Surfaces such as cutting blocks that are subject to scratching and scoring should be resurfaced if they can no longer be cleaned and sanitized effectively. Or, they should be discarded if they are incapable of being resurfaced. Wooden cutting boards may be prohibited in some areas; if used they must be made of hard maple. Cutting boards made of food-grade seamless hard rubber or acrylic are recommended. Consider purchasing separate sets of cutting boards for different types of foods. Color-coded cutting boards are available: red for meat, yellow for poultry; white for dairy; green for fruits and vegetables; beige for cooked poultry; and blue for seafood. Even a two-board combination (such as red for raw foods and beige for cooked foods) will reduce the risk of cross contamination.

Thermometers. Monitoring proper storage, cooking, holding, cooling and reheating temperatures is vital to food protection. To monitor food temperatures, foodservice operations must use metal food thermometers capable of measuring temperatures from 0°F (-18° C) to 220°F (103°C) with an accuracy of ±2°F (1°C). The most common thermometer used in foodservice is the bi-metallic stem type. Bi-metallic stem thermometers are made of two metal strips which are joined together.

The expansion and contraction of these strips move a pointer on a dial face. Bi-metallic thermometers need to be calibrated on a regular basis to ensure accuracy.

Technology has offered some alternatives to bi-metallic thermometers. As more foodservice operations implement HACCP food safety programs, other types of foodservice thermometers are becoming more common.

Thermocouple. Thermocouple probes can measure the surface temperature of food and provide a digital read-out. A digital read-out can be read more accurately and quickly than the traditional bi-metallic stem thermometer. Thermocouples are especially useful with thin foods, such as hamburger patties.

Time/Temperature Indicator Strips. Time/temperature indicator strips use liquid crystals which change color when food enters the hazard zone for food temperatures. Time/temperature indicator strips can be incorporated into food packaging.

Figure 7.3 ■ Calibrating a Thermometer

Slush Method for Calibrating a Bi-Metallic Thermometer

■ Fill an insulated container (like a foam cup) full of potable, crushed ice.

■ Add cold water and stir.

■ Allow time for the mixture to come to a 32°F (0°C) temperature (about 4-5 minutes).

■ Insert a bi-metallic stemmed thermometer into the geothermal center of the cup (away from the bottom and sides).

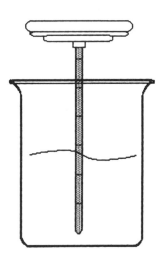

■ Hold thermometer until the temperature stabilizes, and record the temperature.

■ Repeat two times to verify the temperature reading.

■ If the temperature is not 32°F (0°C), use pliers on the calibration nut under the top of the thermometer to adjust the temperature to 32°F (0°C).

Figure 7.4 ■ Measuring a Food Temperature

- Insert into geometric center or into thickest portion.

- Do not touch the sides or bottom of the food container.

- Insert away from the bone, fat or gristle on a food product.

- Insert sideways into thin items such as meat patties.

- Place between frozen packages of food.

- Wait 5-10 seconds for an accurate reading.

- Sanitize after each use.

An important consideration with any type of temperature-measuring device is cross contamination. A thermometer used to measure the temperature of a raw food and then used to monitor a prepared food can contaminate the prepared food with bacteria. Probes should be sanitized with an approved sanitizer after each use. Foodservice storage areas should be equipped with thermometers to measure the ambient temperature. Grill

Figure 7.5 ■ Monitoring Temperature of Food in Storage

Courtesy: Cooper Instrument Corporation

and oven thermometers may be used to verify the calibration of cooking equipment.

Clean-in-Place Equipment. Clean-in-place equipment must be self-draining and designed so that cleaning and sanitizing solutions circulate throughout a fixed system and contact all interior food-contact surfaces and should be self-draining. Clean-in-place equipment that cannot be disassembled for cleaning should have access points to allow inspection of interior food contact surfaces.

Equipment Calibration. Cooking and holding equipment — such as ovens, grills, steam tables and food warmers — should be calibrated on a regular basis. Equipment that is not calibrated may not cook food thoroughly or hold food at a safe temperature.

Figure 7.6 ■ Verifying Calibration of Equipment with a Grill Thermometer

Courtesy: Cooper Instrument Corporation

Mechanical Warewashing Equipment. Dishmachines must be equipped with a temperature measuring device that indicates the temperature of water in each tank and the temperature of water during the sanitizing final rinse. Warewashing machines should be equipped with internal baffles, curtains or other means to minimize internal cross contamination between the solution in the wash and rinse tanks.

Manual Warewashing Equipment. A sink with at least three compartments must be provided for manually washing, rinsing, and sanitizing equipment and utensils. A temperature-measuring device must be provided for frequently measuring the washing and sanitizing temperature of the water.

Equipment Maintenance Responsibilities

Equipment and utensils must be maintained in a state of good repair and condition. Some equipment maintenance responsibilities will undoubtedly be carried out by foodservice staff, while others may be performed by the operation's maintenance staff, or through a contract with a commercial equipment maintenance company. An inspection checklist should be used to ensure that equipment is properly maintained. Foodservice managers should maintain a record of repairs to equipment to assist in justifying repair/replacement decisions. The steps in replacing equipment include:

1) Conducting Cost Analysis: A cost analysis compares the cost of replacing of a piece of equipment with the cost of repairs. A repair log should be maintained to assist in the process. When the cost of materials and labor to repair equipment is near to or greater than the cost of replacement, the equipment should be replaced.

2) Developing Equipment Specifications: When the decision has been made to replace equipment, the foodservice manager should develop a specification for the new equipment. Considerations for developing purchase specifications include:

- Will the equipment provide the time and temperature control that is necessary for safe food?

- Is the equipment properly sized for the volume of food that is prepared?

- Is the equipment reliable or is it prone to frequent breakdowns?

- Is the equipment designed so that it can be easily cleaned and sanitized?

3) Installation and Staff Training: The manufacturer or supplier should assist in ensuring that the new equipment is properly installed and

Figure 7.7 ■ Equipment Maintenance Checklist

Equipment	Clean?		Operational?		Corrective Actions
	Y	N	Y	N	
Oven					
Range Top					
Broiler					
Tilting Skillet					
Meat Slicer					
Mixer					
Food Carts					
Steamer					
Trunnion Kettles					
Steam-Jacketed Kettle					
Food Chopper					
Blender					
Grill					
Fryer					
Steam Table					
Food Warmer					

in providing training to staff. Often, vendors will charge an additional fee for equipment manuals. Foodservice managers and other purchasers should attempt to negotiate the price of equipment operation manuals, staff training, and other support into the purchase price of the equipment.

Key Concepts

■ The equipment used to prepare food and the physical facilities of the foodservice operation have a tremendous impact on the safety of food served.

■ A well designed foodservice operation utilizes materials that are durable, non-porous, and easy to clean.

■ Cleanability is the extent to which an item is accessible for cleaning and inspection and ease in which soil can be removed by normal cleaning methods.

■ Fixtures and equipment used for food preparation, cleaning, or sanitizing may not be directly connected to a sewer connection.

■ A cross-connection can result in backflow and back-siphonage.

■ If an air gap is not present, then the equipment should be equipped with a mechanical device to prevent backflow.

■ Equipment and utensils must be maintained in a state of good repair and condition.

Case Study

Henry, the foodservice manager at Pinewood Correctional Facility, has identified several problems with equipment in the foodservice operation. There are frequent breakdowns of the refrigerator equipment caused by clogged air filters. The Engineering and Maintenance staff are responsible for checking the filters on a routine basis, but they often forget to complete the task. The convection oven frequently fails to heat food on a timely basis and has required frequent repairs. The cutting boards are old and heavily nicked, but there is no money in the current budget to replace them.

☛ 1. What could be done to prevent the refrigeration break-downs?

☛ 2. What is a possible consequence of the oven not working properly?

☛ 3. How could Henry justify purchasing new cutting boards?

Review Questions

1. Floors should NOT be constructed of material that is:

    A. non-absorbent

    B. absorbent

    C. durable

    D. cleanable

2. Walls should be constructed of a material that is:

    A. dark in color

    B. light in color

    C. porous

    D. absorbent

3. Which of the following should NOT be selected as a finish for floors in a food preparation area?

    A. terrazzo

    B. vinyl

    C. quarry tile

    D. sealed concrete

4. A material used for walls that is most resistant to impact from movable carts and equipment is:

    A. terrazzo

    B. structural glazed tile

    C. sealed concrete

    D. ceramic tile

5. A ceiling that is economical and has sound-absorbing qualities is:

   A. drywall

   B. plaster

   C. acoustical tile

   D. exposed concrete

6. A direct link between the sewage system and an equipment drain is called a (an):

   A. backflow

   B. cross-connection

   C. back-siphon

   D. air gap

7. A seal found on equipment that assures the foodservice manager that the item meets industry quality standards has the letters:

   A. FSE

   B. FMP

   C. NIF

   D. NSF

8. Equipment that is on elevated legs should be how many inches off the floor?

   A. 4"

   B. 6"

   C. 8"

   D. 10"

9. Which material would NOT be appropriate for a cutting board?

    A. acrylic

    B. hard rubber

    C. maple

    D. pine

10. What is required at the juncture of floors and walls in food preparation areas?

    A. matting

    B. coving

    C. tile

    D. caulking

Safety Management

Chapter Objectives

■ Identify the role of OSHA in protecting worker safety

■ Discuss the causes and prevention of common foodservice accidents

■ Recognize the importance of accident investigation

■ Review the types of fires and fire extinguishers

■ Discuss compliance with the Hazard Communication Standard

Food safety is concerned with protecting the public's health. In addition to preventing foodborne illness, a foodservice professional must be concerned with preventing accidents and injuries.

Workplace safety is related to food protection and is important to foodservice managers. Every foodservice operation should have a plan in place to prevent accidents.

Safe Work Environment

An accidental injury to a customer may result in legal action against the foodservice operation and an economic loss due to legal costs, increased insurance rates, and loss of business. An employee accident could result in workers' compensation costs, lost time from work, and has a negative impact on morale. Accidents and injuries can be prevented through safety training and the elimination of hazards. Every staff member shares in the responsibility for maintaining a safe work environment. Some general guidelines for safety management include:

■ Inspect work areas to ensure their safety.

■ Maintain equipment in proper working order.

■ Staff appropriately to avoid employee accidents resulting from haste.

■ Train every employee in safe operating and work practices.

■ Investigate all accidents and take corrective action to prevent re-occurrence.

■ Conduct safety meetings and maintain safety records.

■ Recognize employee efforts to prevent accidents and maintain a safe work environment.

Common Foodservice Accidents

Slips, Trips, and Falls. The causes of slips, trips, and falls include wet floors, blockage of narrow walkways by equipment, and staff using crates and shelves instead of ladders.

Burns. Burns are a common accident in foodservice operations. Stoves, ovens, steamers, toasters, and coffee pots are a source of danger for foodservice workers. Foodservice workers must be aware of their work environment and avoid careless actions which may result in burns.

Cuts. Cuts are often the result of either a staff member not paying careful

Figure 8.1 ■ Safety Self-Inspection Checklist

SAFETY INSPECTION	Check ✔		Corrective Actions
	YES	NO	
Tile and carpeting is in good repair			
Spills are wiped up immediately			
"Wet floor" signs are available and are used			
All areas are adequately lighted			
Ladders are inspected and in good repair; staff do not use chairs, crates, or shelves for reaching high objects			
Stairs are not cluttered and handrails are in good repair			
Walkways are not cluttered			
Knives are stored in a holder when not in use			
Guards are in place on all cutting or slicing equipment			
Electrical cords are in good repair			
Equipment such as mesh gloves, goggles, and plastic gloves are accessible and staff are trained in their use			

attention or staff not being trained in safe work habits. Staff should be trained in the use of slicers and food choppers before being permitted to operate them. Mesh gloves and blade guards should be used to reduce the possibility of injury.

Strains and Sprains. Strains and sprains can be the result of improper lifting techniques. Staff should be encouraged to ask for assistance and to use a dolly or hand truck when lifting heavy objects. Staff should also be trained in proper lifting techniques. When lifting heavy objects, staff should remember to bend at the knees and to lift using the strength of the legs rather than bending at the waist and lifting with the back. When lifting, staff should make sure that they have a firm grasp on the object and lift with a smooth action, keeping the object close to their body.

Electrical Shocks. Electrical shocks are often the result of frayed or damage power cords. All electrical equip-

ment should be inspected on a regular basis, and frayed cords must be replaced immediately. Another source of electrical shocks is the splashing of water onto electrical outlets or components. More than an unpleasant experience, an electrical shock could result in cardiac arrest and seriously endanger an employee.

Investigating Injuries and Illnesses. All employees should be required to report any work-related injury or illness to the foodservice manager. An injury is often the result of an accident such as a burn, fall, or cut. An occupational illness may result from exposure to a hazard over time. For instance, a cashier may develop a health problem from making repetitive motions.

For any injury, employees should submit an accident report that documents the nature of the injury or illness, when the accident occurred, and any circumstances surrounding the accident. The foodservice manager must investigate the accident and

Figure 8.2 ■ Safe Work Practices

- Use "wet floor" signs when mopping.

- Do not leave spills on the floor.

- Require staff to wear skid-resistant shoes.

- Do not allow staff to use an unsafe ladder.

- Do ensure adequate lighting, even in walk-in coolers and storage rooms.

- Keep carts, crates, and boxes out of walkways.

- Replace broken floor tiles or worn/ripped carpeting immediately.

- Keep doors and drawers closed.

- Ask for assistance when removing a heavy, hot object from the stove or oven.

- Keep pot handles turned inward, not out toward the walkway.

- Use dry hot pads to remove hot items from the oven.

- Allow steam to escape before reaching into a steamer.

- Vent steam from a corner of the pan facing away from you when removing the lid from items in the oven or on the stove.

Figure 8.2 ■ Safe Work Practices *continued*

- Use a small pan to ladle hot liquids, rather than pouring from a heavy container.

- Remind co-workers about hot surfaces.

- Never try to catch a falling knife.

- Cut away from yourself when using a knife.

- Never use a knife as a can opener.

- Store knives in a holder when not in use.

- Never put a knife in the pot and pan sink. Wash and sanitize individually.

- Train staff in the proper use of knives.

- Wear protective mesh gloves when using or cleaning a slicer or other chopping equipment.

- Operate slicing and chopping equipment only with all of the guards in place.

- Disconnect the electrical power to slicing and chopping equipment before cleaning.

determine the cause. By determining the cause, the foodservice manager can determine what actions to take to prevent the accident from occurring again. Safety problems may be corrected through staff training (e.g. training program on safe lifting techniques) or by correcting hazards in the workplace (e.g. replacing frayed electrical cords).

Fire Safety

A potential for a fire exists in all foodservice operations. Possible causes of fire include: grease build-up on exhaust vents, equipment malfunction, ignition of trash because of cigarettes being improperly discarded or food left unattended during preparation. Employees must be familiar with possible sources and kinds of fire, the locations of fire suppression equipment and the actions to take in the event of a fire.

Fire Extinguishers. For responding to a fire, the first few minutes are the most important. A fire requires three conditions: fuel, oxygen, and heat. The goal in extinguishing a fire is to remove one or more of the required elements. Fires are classified according to the material combusting, as shown in Figure 8.3. Extinguishing Class B and Class C fires generally requires the

---

Figure 8.3 ■ Types of Fires and Fire Extinguishers

Class	Type of Fire	Recommended Extinguisher
Class A	Normal combustible materials	Foam, soda acid, pump tank (plain water), gas cartridge, multipurpose dry chemical
Class B	Grease and oil	Foam, carbon dioxide, multipurpose dry chemical, ordinary dry chemical
Class C	Electrical	Carbon dioxide, multipurpose dry chemical, ordinary dry chemical

Figure 8.4 ■ Accident Investigation Form

Form 100
Rev. 1-82

OSHA
Case or File No. _____

DEPARTMENT OF LABOR
Bureau of Worker's Disability Compensation
LANSING, MICHIGAN
EMPLOYER'S BASIC REPORT OF INJURY

COPIES TO BE DISTRIBUTED
Yellow and Green – Bureau of Workers's
Disability Compensation, Lansing, Mich.
Blue – Insurance Company
Pink – Employer File
White – Employee

Employers must report to the Bureau on Form 100 all injuries, including diseases, which arise out of an in the course of the employment and cause: 1. Seven (7) or more days of disability not including Sundays or the day of injury. 2. Death. 3. Specific Losses. In case of DEATH also file immediately an additional report on Form 106.

1. **INJURED EMPLOYEE**_____ Soc. Sec. No. _____/_____/_____

2. Address _____ Telephone No. _____

3. Birthdate - Month _____ Day _____ Year _____   If under 18, date working permit issued _____

4. Sex: ☐ Male ☐ Female   Number of injured employee's children under age 16 living with injured _____

5. Marital status: ☐ Married ☐ Single   If married male, is wife living with him? ☐ Yes ☐ No

6. Number of other family members or relatives at least 50% supported by injured _____

7. **DATE OF INJURY** _____ Last day worked _____ Did employee die? ☐ Yes ☐ No   If yes, date _____

8. Location of Injury   City _____ State _____ County _____

9. Was place of accident or exposure on employer's premises? ☐ Yes ☐ No

10. Name and address of physician _____

11. If hospitalized, name and address of hospital _____

12. DESCRIPTION OF ALLEGED INJURY (Complete and specific information needed for each category)

A. Describe the injury or illness
   Examples: Amputation, Burn, Cut, Fracture, Sprain, etc.

B. Part of body - The part of body directly affected by the injury or illness.
   Examples: Head, Arm, Leg, Circulatory system, etc.

C. Describe the events that caused the injury. Examples: Fell, Operating machinery, Exposure to chemicals, etc.

D. Name the object or substance which directly injured the employee.
   Examples: Knife, Band Saw, Acid, Floor, Oil, Punch Press, etc.

13. Occupation of injured employee (be specific) _____

14. Department _____ Foreman or supervisor _____

15. Total Gross Wages - Highest 39 of 52 weeks preceding date of injury
   Total Gross Weekly Wages $ _____ No. of weeks used in calculation _____
   Average weekly wage $ _____

16. Complete the following only if the injured employee received wages from a second employer.
   Name of second employer _____
   Mailing address _____

17. Date returned to work _____ or estimated lost time from work _____

18. IS EMPLOYEE CERTIFIED AS VOCATIONALLY HANDICAPPED?   ☐ Yes ☐ No

19. IS EMPLOYEE RECEIVING UNEMPLOYMENT INSURANCE BENEFITS?   ☐ Yes ☐ No

20. **EMPLOYER**   MESC. No. _____
   A. _____ Federal ID No. _____
   B. _____

21. Location (if different from mail address) _____

22. **TYPE OF BUSINESS** _____

23. **INSURANCE COMPANY (Not agent)**   Carrier ID No. _____

24. **HAS WHITE COPY OF THIS REPORT BEEN GIVEN TO EMPLOYEE?**   ☐ Yes ☐ No

Questions or errors should be immediately reported to the employer representative indicated below

Date of Report _____ Prepared by _____

Signature (in ink) **Employer** or Representative   Tele. #

Figure 8.5 ■ Fire Safety Self-Inspection Checklist

FIRE SAFETY INSPECTION	Check ✔ YES	NO	Corrective Actions
Hoods and vents are clean and free of grease			
Power cords are not frayed or damaged			
Broiler, salamander, grill, fryer and range top are clean and free of grease build up			
Smoking is limited to designated areas			
Cigarette butts are being properly disposed of			
Combustible materials, such as paper supplies, are properly stored — away from heat source			
Staff are following safe work practices, e.g. hot pads are kept away from heat source			
Preventive maintenance on electrical equipment is up-to-date			
Fire extinguishers are charged and operable			
Smoke detectors are functioning properly			
Supplies are not stored in a manner that would interfere with the sprinkler system			

removal of oxygen, while a Class A fire may require removing the heat or the fuel. Fire extinguishers have an identification system that indicates which type of fire they will extinguish. The best all-purpose fire extinguisher for a foodservice operation is the multi-purpose dry chemical type. Extinguishers should be periodically inspected to ensure that they are still properly charged.

Hazard Communication

Occupational Safety and Health Administration (OSHA). OSHA is a government agency responsible for protecting the safety of workers, including those in foodservice establishments. In 1971, the Occupational Safety and Health Act was enacted. To enforce the act, OSHA was formed in the same year. OSHA develops and enforces mandatory regulations and standards to ensure a safe work environment. OSHA requires the maintenance of safety records, the filing of safety reports, and strict adherence

to good safety practices. Employers are required to maintain records of employee accidents and to report serious accidents and fatalities to OSHA.

To monitor compliance, OSHA periodically inspects businesses to determine the level of compliance. If a business is found to be in violation of a health and safety regulation, OSHA may issue a citation, fine the employer, or in some cases order the employer to close the business until compliance is attained. OSHA also provides training information for employers and workers. Employers are required to display and keep displayed a poster prepared by the Department of Labor summarizing the major provisions of the Occupational Safety and Health Act and telling employees how to file a complaint. The poster must be displayed in a conspicuous place where employees and applicants for employment can see it.

OSHA has issued a regulation requiring employers to inform workers about the possible dangers of chemicals they use to do their jobs, and to train them to use chemicals in a safe manner. The regulation is called the *Hazard Communication Standard*. The purpose of Hazard Communication Standard is : 1) to ensure that all employers receive the information they need to inform, 2) to train their employees properly on the hazardous substances they work with, and 3) to help design and put in place employee protection programs. It also provides necessary hazard information to employees, so they can participate in and support the protective measures in place at their workplaces. To comply with the Hazard Communication Standard:

■ Review the Hazard Communication Standard.

Figure 8.6 ■ Hazard Communication Training Program Content

■ Location of Material Safety Data Sheets (MSDS) in the operation

■ Inventory list of hazardous chemicals used in the operation

■ How to read product labels to learn potential hazards and directions for safe use

■ Proper handling, application, storage and disposal of chemicals

■ Use and location of personal protective equipment

■ Emergency procedures in the event of a spill or exposure

Figure 8.7 ■ Sample Material Safety Data Sheet

Material Safety Data Sheet
May be used to comply with OSHA's
Hazard Communication Standard, 29 CFR
1910 1200. Standard must be consulted for
specific requirements.

U.S. Department of Labor
Occupational Safety and Health
Administration
(Non-Mandatory Form)
Form Approved
OMB No. 1218-0072

IDENTITY *(as Used on Label and List)*	*Note: Blank spaces are not permitted. If any item is not applicable or no information is available, the space must be marked to indicate that.*

Section I

Manufacturer's name	Emergency Telephone Number
Address (Number, Street, City, State and ZIP Code)	Telephone Number for Information
	Date Prepared
	Signature of Preparer *(optional)*

Section II—Hazardous Ingredients/Identity Information

Hazardous Components (Specific Chemical Identity, Common Name(s))	OSHA PEL	ACGIH TLV	Other Limits Recommended	% (optional)

Section III—Physical/Chemical Characteristics

Boiling Point		Specific Gravity (H_2O = 1)	
Vapor Pressure (mm Hg)		Melting Point	
Vapor Density (AIR = 1)		Evaporation Rate (Butyl Acetate = 1)	
Solubility in Water			
Appearance and Odor			

■ Develop an inventory of chemicals used in the foodservice operation.

■ Request a Material Safety Data Sheet (MSDS) for each chemical.

■ Put the MSDS in a location available to employees.

■ Make sure that all chemicals are in labeled containers.

■ Include safe chemical use in the new employee orientation and as part of annual training for all employees.

Key Concepts

■ A foodservice professional must be concerned with preventing accidents and injuries.

■ Accidents and injuries can be prevented through safety training and the elimination of hazards.

■ Every staff member shares in the responsibility for maintaining a safe work environment.

■ OSHA is a government agency responsible for protecting the safety of workers, including those in foodservice establishments.

■ Common foodservice accidents include slips, falls, trips, cuts, burns, strains, sprains, and electrical shocks.

■ The foodservice manager should investigate all accidents to determine the cause.

■ A potential for a fire exists in all foodservice operations.

■ The Hazard Communication Standard requires employers to inform workers about the possible dangers of chemicals they use to do their jobs and to train them to use chemicals in a safe manner.

Case Study

Juan is a member of the wait staff at Highlands Retirement Center. He wears a uniform and dress shoes at work. Clearing tables during a busy time, Juan takes dirty dishes into the dishroom. As he leaves the dishroom in a hurry, he slips and falls onto water spilled near the dishroom entrance. During your investigation, you will need to examine factors that contributed to the accident and identify how to prevent future accidents.

☞ 1. What were the major causes of this accident?

☞ 2. How could this accident have been prevented?

☞ 3. What would you do to prevent future accidents?

Review Questions

1. An agency that is responsible for enforcing workplace safety standards is:

   A. USDA

   B. OSHA

   C. FDA

   D. CDC

2. A form that describes the proper use, precautions, ingredients and emergency procedures for a chemical is called a:

   A. Safe Use and Precaution Sheet

   B. Chemical Safety Data Sheet

   C. Occupational Safety Data Sheet

   D. Material Safety Data Sheet

3. A fire resulting from the ignition of grease build-up would be classified as:

   A. Class A

   B. Class B

   C. Class C

   D. Class D

4. The best type of all-purpose fire extinguisher for foodservice operations is:

   A. foam

   B. ordinary dry chemical

   C. multi-purpose dry chemical

   D. carbon dioxide

5. Which of the following is NOT a requirement for a fire?

    A. hydrogen

    B. oxygen

    C. heat

    D. fuel

6. Which is the following is NOT a technique for preventing cuts?

    A. operate slicing equipment with guards in place

    B. store knives in a holder

    C. cut towards yourself carefully

    D. using a dust pan to collect broken glass fragments

7. Which of the following is NOT required to be related as part of a Hazard Communication Program?

    A. location of MSDS

    B. cost of chemical supplies

    C. use of personal protective equipment

    D. emergency procedures

8. A burn is an example of a(n):

    A. injury

    B. illness

    C. hazard

    D. risk

9. To safely clean and sanitize a chef's knife:

    A. put it in the pot and pan sink and manually wash with other items

    B. clean and sanitize individually in the pot and pan sink

    C. rinse thoroughly in a food preparation sink and store in a knife holder

    D. rinse with a sanitizing solution and store in a utensil drawer

10. Which of the following is NOT a safety hazard?

    A. pot handles pointed toward the center of the range top

    B. carefully using a metal crate as a step to reach the top shelf in the cooler

    C. using a dry hand towel to remove items from the oven

    D. prying frozen chicken apart with a sharp knife

Regulations, Inspections, and Crisis Management

Chapter Objectives

- Identify regulatory agencies and resources related to food protection

- Describe the inspection process

- Identify ways to prepare for inspections

- Provide examples of crisis planning

- Identify how to handle a complaint of foodborne illness

- Discuss considerations for communication during a crisis situation

Foodservice operations must comply with many local, state, and in some cases Federal regulations. Understanding regulations and standards is the first step in complying successfully. Many regulatory agencies are also a source of assistance for foodservice managers. In the event of a crisis, the foodservice manager will interact with these same agencies, so information on crisis management is included in this chapter.

Regulations

Each of the three branches of the government is involved in protecting the safety of the food supply and public health. Congress and the Senate are the legislative branch, and they pass laws regarding food handling and protection. Agencies with the executive branch of the government develop regulations to implement the laws, and the judicial branch reviews the policies and procedures of regulatory agencies to protect the rights of both private businesses and the public. Laws usually state broad objectives and need to be interpreted. Regulations are more specific and give guidance in how to comply with the law. Several regulations may be necessary to implement one law.

A foodservice operation may be required to comply with a number of different laws and regulations regarding safety and sanitation. Laws and regulations regarding food protection are present at the Federal, state and local level. It is important to the food-service manager to learn about standards and expectations of different regulatory agencies and to develop policies and procedures to achieve compliance. Different agencies may have different expectations. For instance, local laws may be stricter than state laws. In this case the foodservice operator must comply with the highest standard. When in doubt, the foodservice operator should consult a representative of the regulatory agency.

Local or Municipal Laws and Regulations. Depending on the size of the community and the number of foodservice operations it has, the county or municipal government may have a department of health or food protection that enforces local laws.

State Laws or Regulations. All states have a department charged with protecting the safety of food produced, served, and transported in the state. The office may be within in the department of public health or the department of agriculture.

Federal Laws and Regulations. Federal laws and regulations exist to protect food safety by requiring inspection, controlling food additives, and regulating the transport of food across state borders.

Regulatory Agencies and Resources

Food and Drug Administration (FDA). The FDA, an agency of the Department of Health and Human Services' Public Health Service, is responsible for ensuring the safety and wholesomeness of all foods sold in interstate commerce except for meat, poultry and eggs, all of which are under USDA jurisdiction. FDA develops standards for the composition, quality, nutrition and safety of foods, including food and color additives. It does research to improve detection and prevention of food contamination. It collects and interprets data on nutrition, food additives and environmental factors, such as pesticides, that affect foods. FDA also sets standards for certain foods and enforces Federal regulations for labeling, food and color additives, food sanitation and safety of foods. FDA monitors recalls of unsafe or contaminated foods and can seize illegally marketed foods.

U.S. Department of Agriculture (USDA). Through inspection and grading, the USDA enforces standards for wholesomeness and quality of meat, poultry and eggs produced in the United States. USDA is involved in nutrition research and in educating the public about how to choose and cook foods and how to manage healthy or restricted diets. USDA food safety activities include inspecting poultry, eggs, and domestic and imported meat; inspecting livestock and production plants; and making quality (grading) inspections for grain, fruits, vegetables, meat, poultry and dairy products (including Brie and other cheeses). USDA's education programs target family nutritional needs, food safety, and expanding scientific knowledge. The department supports education with grants in food and agricultural sciences and conducts its own cooperative food research.

Centers for Disease Control and Prevention (CDC). An agency of the Department of Health and Human Services, CDC becomes involved as a protector of food safety — including responding to emergencies when foodborne diseases are a factor. CDC surveys and studies environmental health problems. It directs and enforces quarantines, and it administers national programs for prevention and control of vector-borne diseases (diseases transmitted by a host organism) and other preventable conditions.

National Marine Fisheries Service (NMFS). A part of the Department of Commerce, NMFS is responsible for seafood quality and identification, fisheries management and development, habitat conservation, and aquaculture production. NMFS has a voluntary inspection program for fish products. Its guidelines closely match

regulations for which FDA has enforcement authority.

State and Local Governments. State and local government agencies cooperate with the Federal government to ensure the quality and safety of food produced within their jurisdictions. FDA and other Federal agencies help state and local governments develop uniform food safety standards and regulations, and assist them with research and information. States inspect restaurants, retail food establishments, dairies, grain mills, and other food establishments within their borders. In many instances, they can embargo illegal food products, which gives them authority over fish, including shellfish, taken from their waters. FDA provides guidelines to the states for this regulation. Twenty-eight states have their own fish inspection programs. FDA also provides guidelines for state and local governments for regulation of dairy products and restaurant foods. The departments responsible for food safety and inspection functions vary by state and community. Some are divisions of other agencies, such as state agriculture or health departments.

Model Food Code

The Model Food Code is a framework upon which an effective retail food safety program can be built. Today, FDA's purpose in maintaining an updated model food code is to assist food control jurisdictions at all levels of government by providing them with a scientifically sound technical and legal basis for regulating the retail segment of the food industry. The retail segment includes those establishments or locations in the food distribution chain where the consumer takes possession of the food.

The Model Food Code is neither Federal law nor Federal regulation, and it does not replace local regulations. Rather, it represents FDA's best advice for a uniform system of regulation to ensure that retail food is safe and properly protected and presented. Although not Federal requirements

(until adopted by Federal bodies for use within Federal jurisdictions), the Model Food Code provisions are designed to be consistent with Federal food laws and regulations, and are written for ease of legal adoption at all levels of government. Providing model food codes and model code interpretations and opinions is the mechanism through which FDA promotes uniform implementation of national food regulatory policy among the several thousand Federal, state, and local agencies.

This book utilizes the 1997 Model Food Code as one of its references on proper food temperatures and food handling practices.

Variances. Regulatory agencies may allow variances or exceptions to regulations or codes in some cases. For example, the FDA Model Food Code requires that cold food be stored at 41° F (5°C) or below. But the FDA will accept food being held at 45°F (7°C) if the existing refrigeration equipment is not capable of maintaining the food at 41°F (5°C), and if the equipment is upgraded or replaced within five years to maintain food at a temperature of 41°F (5°C).

Inspections

One way agencies and governments assess compliance with regulations is through the inspection process. The FDA recommends that foodservice operations be inspected once every six months. An inspection often begins with an inspector or sanitarian identifying him or herself and presenting the appropriate identification. If an inspector does not present identification, be sure to ask for it. In most cases inspections are unannounced. In some areas of the country, inspections only occur in response to complaints. In other areas, routine inspections can be expected by every foodservice operation. Some foodservice operations may be inspected by several different agencies. For instance, healthcare organizations may be inspected by the local department of health, the Health Care Financing Administration (HCFA) and the Joint Commission on the Accreditation of Healthcare Organizations (JCAHO).

Inspection Process. When interacting with inspectors or surveyors, remember to show professional courtesy. If possible, accompany the inspector during the visit. This way you will be available for questions that the inspector may have, and you can ask the inspector questions if you do not understand a regulation or violation. The key to success with inspections is to view them as a source of help and not a hindrance. Inspectors are well trained in the area of food protection and are a source of valuable information.

Areas that an inspector will probably address include:

■ **Food Storage Practices,** including the cleanliness and temperature of food storage areas; indication that proper stock rotation methods (FIFO) are being used; labeling and dating practices for food products; protection of food in storage from cross contamination; and proper storage of chemicals away from food preparation areas, and in clearly marked containers.

■ **Food Preparation and Service Practices,** including the thawing, cooking, holding, transporting, serving, cooling and reheating of food. The inspector will want to review temperature logs, record temperature measurements, and observe staff work habits.

■ **Foodservice Personnel Hygiene,** including the cleanliness of staff, use of disposable gloves and utensils to avoid hand contact with food, presence of any staff with a transmittable disease, accessibility and use of hand wash sinks, hand wash supplies and toilet facilities. The inspector may interview staff about work habits and practices.

■ **Cleaning and Sanitization Practices,** including manual and mechanical warewashing practices, proper dilution of chemical sanitizers, and proper water temperature. The inspector will tour the premises and inspect equipment to determine overall cleanliness. The inspector may request a copy of cleaning schedules.

■ **Facilities,** including safe water supply, proper sewage disposal and backflow prevention with air gap or other approved device, presence of pest infestations, and pest control methods and techniques employed.

Have food temperature logs and staff training records available for the inspector. Some facilities are required to have a manager certified in foodservice sanitation on the premises at all times. Be sure that the proper records and evidence of certification are available and current.

If the inspector discovers minor infractions, he or she will establish a deadline for correction. Correct violations immediately — even before the inspector leaves the premises if possible. Take notes during inspection. The notes will help you to make any necessary improvements and are a resource to prepare for future inspections. If a serious problem or imminent health hazard is noticed, the inspector may order the facility to close immediately. A foodservice operation that fails to correct problems by the deadline risks fines and loss of the permit to operate a foodservice operation.

HACCP-Based Inspections. Some local health departments have changed their inspection process to one based on HACCP principles. Instead of traditional inspections which focus on the foodservice operation, a HACCP-based inspection focuses on food. A HACCP-based inspection might take several days as inspectors follow food through the operation. Inspectors measure and observe critical limits at each step in the flow of food. In traditional inspections, ratings are on a scale from 0 to 100, with 100 being a perfect score. Points are de-

ducted for each violation. In HACCP-based inspections, points are assigned for each violation and 0 is a perfect score. In a traditional inspection, you would be penalized once for a refrigerator that was not in the proper temperature range. In a HACCP-based inspection, you may be penalized for each food in the refrigerator that was at an unsafe temperature.

The inspector will also review HACCP logs during the visit. Reviewing the logs allows the inspector to inspect food safety practices over a long period of time rather than just on the day of the inspection.

Inspection Preparation. An inspector will usually leave a written report which notes areas for improvement and any violations. Study the written report to identify why violations occurred and to develop an action plan to avoid having the same violation occur a second time. To improve compliance during inspections, follow these guidelines:

■ **Plan and Prepare:** Obtain a copy of applicable regulations and review them in advance. If you have questions about the regulations, contact the regulatory agency that enforces them. Develop policies and work practices that comply with the regulations.

■ **Self-Inspection:** Conduct routine self-inspections to determine strengths and weaknesses of the facility. Ask an inspector for blank inspection reports, or create self-inspection forms specifically for your foodservice operation. Several examples of self-inspection forms are included in this textbook. Correct problems that you discover during self-inspections.

■ **Network:** Through professional organizations, discuss regulations and compliance issues with peers. Identify common strategies and resources.

■ **Train and Educate Staff:** Train all staff members in safe food preparation practices and regulations applicable to their own jobs.

Crisis Management

A crisis is a sudden change, often an unstable condition, that requires decisive action to be taken. Often a crisis endangers the health of a customer or employee, or it threatens the safety and security of the foodservice operation. A crisis can occur in any type of foodservice operation. Some examples of crisis situations are:

■ Fire

■ Flood

■ Tornado or hurricane

■ Utility (water or gas) loss

■ Delivery driver strike

■ Food tampering by customer or employee

■ Sudden illness of an employee or customer

■ Workplace violence

■ Bomb threat

■ Robbery or theft

■ Vandalism

■ Act of terrorism

■ Foodborne illness outbreak

Crisis Management Plan. Imagine being the manager of a foodservice operation that has been flooded and is without utilities. What if the foodservice operation were at a hospital, nursing home or other facility where you were expected to continue providing meals. How would you do it? How could you be sure that the food being served is safe?

To properly handle such a situation, it is necessary to pre-plan. Anticipate the type of crisis situations that your operation might face and develop a plan of action. Examples of pre-planning and preparation for crisis situations are:

■ A nursing home maintains a supply of bottled water and a one-week supply of disposable dishes for use in the event the water supply becomes contaminated or service is lost.

■ A fire evacuation plan is established for a fine dining restaurant and employees are trained in the use of fire extinguishers.

■ Employees at a hotel are trained in procedures to follow if they receive a bomb threat, such as what to listen for and what to ask the caller.

■ A fast food restaurant has an established procedure to follow if they receive a complaint about foodborne illness.

■ Emergency phone numbers for police, fire, and medical personnel are posted by each telephone in a college dining hall.

■ Employees of a correctional facility are trained in non-violent intervention techniques.

■ A resort maintains a crisis management manual with procedures to follow in the event of most emergency situations.

Foodborne Illness Outbreak. A foodservice operation may at some time receive a complaint about foodborne illness. Every foodservice operation should have a plan in place to deal with such complaints. Questions that should be asked of the complainant include:

■ What is your name, address and phone number?

■ When did you eat at the foodservice operation (day and time)?

■ What were you served?

■ When did you become ill?

■ Have you sought treatment? Where?

■ Was anyone else with you when you ate at the foodservice operation?

■ If yes, are they ill and what did they have to eat?

After receiving the complaint, contact the local department of health. Reporting the complaint to local public health officials should be done quickly to prevent the illness from spreading further. If the suspect food is still in service, immediately take it out of service. Save the suspect food. A laboratory analysis can help to confirm the presence of foodborne illness or establish the safety of the food item. Do not accept responsibility for the illness or agree to pay for medical care, but encourage the complainant to seek medical care if they have not already done so. After securing the food sample and reporting the complaint to the health department, you may wish to contact your liability insurance carrier and to interview staff involved in the preparation of the suspect food.

The local health department will investigate the complaint of foodborne illness. One way to prevent future outbreaks is to learn from past mistakes. The goals of an investigation are to:

■ Identify which meal and specific menu item caused the illness

■ Identify the organism or other hazard which caused the outbreak

■ Determine the source of the contamination

■ Determine whether any errors or mishandling occurred that allowed the risk of contamination from the original source to increase.

To accomplish this, a team of investigators will interview the victim, visit the foodservice operation to interview management and personnel, take samples of suspected food items, and evaluate the food protection practices of the foodservice operation. HACCP logs can provide a track record of food handling practices in the establishment. Since foodservice workers may be carriers of an illness, the team may request that cultures be taken from all staff to detect the presence of a communicable disease. Bacterial sampling may also be done of work surfaces and equipment to determine if it is contaminated. During an investigation, it is very important for management and personnel to cooperate fully.

Crisis Communications. In the event of a crisis situation, there should be one designated person responsible for contact with members of the media. All other staff should decline answering questions and refer members of the press to the designated spokesperson. Before answering any questions, the spokesperson should determine the facts surrounding the emergency or crisis situation. In many cases, the spokesperson should consult with legal counsel before answering questions if possible. During an emergency, spokespersons are also encouraged to follow these do's and don'ts:

■ Do provide factual information.

■ Don't speculate or guess.

■ Do develop a prepared statement of exactly what information you want to release and then follow your statement.

■ Don't feel obligated to answer all of the questions you are asked.

■ Do communicate in a clear and concise manner.

■ Don't use jargon or lurid descriptions.

■ Do ask technical experts to provide explanations of complex situations if necessary.

■ Don't allow pointed questions to alarm you.

■ Do inform the media that public health and safety is your priority and that you are cooperating with regulatory agencies.

■ Don't appear uncooperative or intimidated.

Key Concepts

- Foodservice operations must comply with many local, state, and in some cases Federal regulations.

- Each of the three branches of the government is involved in protecting the safety of the food supply and public health.

- A foodservice operation may be required to comply with a number of different laws and regulations regarding safety and sanitation.

- The FDA recommends that foodservice operations be inspected once every six months.

- Inspectors are well trained in the area of food protection and are a source of valuable information.

- Correct violations immediately — even before the inspector leaves the premises if possible.

- Conduct routine self-inspections to determine strengths and weaknesses of the facility.

- Train all staff members in safe food preparation practices and regulations applicable to their own jobs.

- Effective crisis management depends on planning and preparation.

Case Study

Lynn is the new foodservice manager for State University Dining Services. She is responsible for several dining halls on a large campus. The local health department is conducting a bi-annual inspection of the foodservice facilities on the campus. Lynn is nervous about her first inspection as a manager and has been trying to prepare her staff. The inspector notices a recent delivery stored on the floor in one of the storage rooms and also points out that the garbage cans are dirty.

☛ 1. How could Lynn have prepared for the inspection?

☛ 2. What should Lynn do about the problems identified by the inspector?

☛ 3. How can Lynn prepare staff for the next inspection to avoid repeating the same violations?

Review Questions

1. The agency involved in the inspection of meat, poultry and eggs is the:

    A. FDA

    B. CDC

    C. USDA

    D. NMFS

2. The agency that sets standards for certain foods and enforces regulations on labeling, food and color additives is the:

    A. FDA

    B. CDC

    C. USDA

    D. NMFS

3. The agency that investigates outbreak of diseases, including foodborne illness, is:

    A. FDA

    B. CDC

    C. USDA

    D. NMFS

4. Most food establishments are inspected by:

    A. FDA

    B. USDA

    C. CDC

    D. state and local government

5. Proper management of crisis situations begins with:

    A. contacting authorities

    B. staff training

    C. planning

    D. media relations

6. Many state and local regulations are based on the:

    A. Meat Act

    B. Food and Drug Act

    C. Model Food Code

    D. US Code

7. The FDA recommends inspections of foodservice operations every:

    A. 4 months

    B. 6 months

    C. 1 year

    D. 2 years

8. The first step to take after receiving a complaint of foodborne illness is to:

    A. remove the suspect food from service

    B. notify your organization's attorney

    C. contact the FDA

    D. contact the CDC

9. When dealing with the media following a foodborne illness complaint, you should:

    A. deny the possibility of an outbreak in your operation

    B. express your cooperation with appropriate authorities

    C. threaten the media to leave your premises or else

    D. answer all questions that the media ask of you

10. Which of the following is NOT an acceptable way of preparing for inspections?

    A. conduct self-inspections

    B. train staff in regulatory standards

    C. attend continuing education programs on trends

    D. make improvements on the day of inspection only

Appendix A
Common Food Protection Terms

Aerobe

A type of bacteria that grows best in the presence of oxygen

Air Gap

An unobstructed air space between an outlet of drinkable water and the flood rim set up to prevent cross-connection

Anaerobe

A type of bacteria that grows best in the absence of oxygen

Approved sources

Reputable vendors of commercial food products

Backflow

The flow of contaminated water into potable water

Back-siphonage

The flow of unsafe water into a safe water supply caused by a lower pressure in the safe water supply and resulting in contamination

Bacteria   A type of microorganism that is responsible for most foodborne illnesses

Biological hazards   Include bacteria, viruses, parasites, and fungi; one of the greatest threats to food safety

CAP   Controlled Atmosphere Packaged — a type of reduced-oxygen packaging that extends shelf life

Carrier   A person who harbors organisms causing a communicable disease but who does not present any symptoms of the illness

Chemical hazards   Include pesticides that are sprayed on food, preservatives used to maintain food, toxic metals in cooking equipment, and chemical cleaning materials used in foodservice operations

Clean   Free of visible soil

Clean-in-place (CIP) equipment Must be designed so that cleaning and sanitizing solutions circulate throughout a fixed system and contact all interior food contact services and is self-draining. It does not include equipment such as slicers or mixers that require in-place manual cleaning.

Cleaning agent   A chemical used for dirt-removing. There are four types alkaline detergents; abrasive cleaners, acid cleaners, and degreasers.

Cleaning schedule

An established schedule for the cleaning needs of a facility. It contains WHAT gets cleaned, WHEN it is cleaned, HOW to clean it and WHO is responsible for the cleaning

Crisis

A sudden change, often an unstable condition, that requires decisive action to be taken

Crisis management

The preparation/practice of a plan that allows a facility to fulfill contractual obligations to the client in the safest possible way during a crisis

Critical control point

A process or step of a process at or by which a preventive measure can be exercised that will eliminate, prevent, or minimize a hazard that occurred prior to that point

Critical limit

The maximum or minimum value to which a physical, biological, or chemical parameter must be controlled at a critical control point

Cross-connection

A dangerous link between an outlet of a drinkable water system and unsafe water or chemicals

Cross contamination

The transfer of harmful microorganisms to food. It can occur in many ways, including contact from human hands, the use of unsanitary equipment or work surfaces, from the storage of raw foods above ready-to-eat foods, and the use of unsanitary cleaning cloths.

Facultative
A type of bacteria that can grow with or without the presence of oxygen

FDA Food Code
A compilation of model requirements for safeguarding public health and ensuring food is unadulterated and honestly presented when offered to the consumer. It is offered by the FDA for adoption by local, state, and Federal government jurisdictions that have responsibilities for foodservice, retail food stores, or food vending operations.

FIFO
An inventory management method which means first in, first out. FIFO helps to ensure that the oldest food products are used first.

Foodborne illness
A disease that is carried or transmitted to people from food

Foodborne illness outbreak
An incident in which two or more people experience the same illness after eating a common food and laboratory analysis verifies that the food is the source of the illness

Foodborne infection
A disease that results from eating food containing living microorganisms

Foodborne intoxication
A disease that results from eating food containing toxins produced by bacteria, molds, or certain animals or plants

Grades

Classifications of foods by a descriptive term (e.g. *choice*) or number (e.g. *#2*) to ensure uniform quality and give an indication of desirableness

HACCP

Hazard Analysis Critical Control Point, a preventive approach to food safety that involves identifying potential hazards, establishing preventive or control measures, and ongoing monitoring to ensure that standards or critical limits are met

Hazard

An unacceptable contamination, unacceptable growth, or unacceptable survival of harmful microorganisms. A hazard may be physical, chemical, or biological.

Hazard zone

The temperature range in which harmful microorganisms grow most rapidly (41° - 140° F); may also be called *danger zone*

Hygiene

Refers to the cleanliness of one's body. Good personal hygiene is required of all foodservice workers.

IPM

Integrated pest management, an approach to managing pests that includes restricting access, reducing food sources, and working with an licensed pest control operator

Manual dishwashing

Cleaning and sanitizing of pots, pans, dishes and flatware that is generally done with a 3 or 4-compartment sink

MAP   Modified Atmosphere Packaged, MAP foods are packaged in a reduced-oxygen environment in order to extend shelf life

Molds   Microscopically small plants that often appear as fuzzy, colored patches on food. Some molds are toxic.

MSD Sheets   Material Safety Data Sheets contain important safety information about hazardous chemicals. MSD sheets must be available for every chemical found in the work place; they are available from the chemical vendor or manufacturer.

NSF   A seal of approval that is a recognized standard of acceptance for pieces of commercial equipment

Parasites   Organisms that live in another organism (host), from which they obtain nutrients

Pathogen   A disease-causing microorganism, which may be foodborne

Personal hygiene   Practices of cleanliness or personal care habits to maintain health

Pest Control Operator   A person licensed to use chemicals in the control of pests

pH   Expression of the acidity or alkalinity of a substance

Physical hazards

Foreign materials that enter accidentally into food; examples are glass fragments, metal shavings, staples from produce crates, and cigarette particles

Potable

Water that is free of contaminants and safe to drink

Potentially hazardous food

Any food that requires temperature control because it can support the growth of dangerous microorganisms

POTTWA

An acronym for the growth conditions of bacteria: Potentially Hazardous Food, Oxygen, Time, Temperature, Water, and Acidity

Sanitary

Free of harmful levels of microorganisms

Sanitizing agents

Chemicals (iodine, chlorine or quaternary ammonium compounds) or hot water (170-180° F/77-82° C) that reduce the number of harmful contaminants to a safe level

Spores

Thick-walled cell bodies formed by some microorganisms, such as molds and certain bacteria. Spores are resistant to heat and other unfavorable conditions.

Thermocouple

A temperature-measuring device that measures the surface temperature of food and uses a digital display to provide a more accurate reading

Toxin

Poisonous compounds; some are man-made while others may naturally occur in food.

TTI

Time/Temperature Indicator, a temperature-measuring device that uses liquid crystals and may be incorporated into food packaging. It changes color to indicate that food has been held at an unsafe temperature.

UHT

A type of food processing

UL

A seal of approval from Underwriters' Laboratories that indicates compliance of the equipment with electrical safety standards

Virus

Very small pathogens that multiply within the living cells of their host; they do not multiply in food but may be transmitted by it.

Yeast

Microorganisms that multiply by budding and may cause the spoilage of food by fermenting sugar to create alcohol and carbon dioxide

Appendix B
Suggested Responses to Case Studies

Chapter 1

One possible cause of the foodborne illness outbreak is that the cook incorrectly thawed the ground beef. Thawing at room temperature is not an accepted method of thawing. The cook could have thawed the ground beef in the microwave and then continued cooking to a safe internal temperature. The cook did not wash her hands after handling the ground beef and before chopping the ingredients for the tossed salad. The cook should not have handled a ready-to-eat food such as salad with her bare hands. Finally, the knife that was used to open the ground beef package was also used to cut fruit. The outbreak could have been prevented by thawing the beef under refrigeration, washing hands after handling raw food, and using a sanitized knife. The possible consequence is that the children, who are a high-risk group, could become quite ill from foodborne illness.

Chapter 2

Fresh beef steaks should be received at 41°F (5°C) or below. The appropriate corrective action to take is to reject the delivery of beef steaks. Other items on the order should be checked, and if the temperature is not acceptable, they should be rejected also.

Chapter 3

Pat was correct in labeling, dating and immediately storing the leftover tomato sauce. But the pans used were too deep. The pans for a thick food such as meat sauce should be 2" deep or less. Pat could also have used an ice bath to quickly cool the large quantity of leftover sauce.

Chapter 4

Ingredients for Hunan Pork with Vegetables

Pork loin, cubed

Green peppers

Carrots

Onions

Soy sauce

Salt

Water chestnuts, canned

Bamboo shoots, canned

Ginger, fresh

Garlic, fresh

Oil

Cayenne pepper, dried

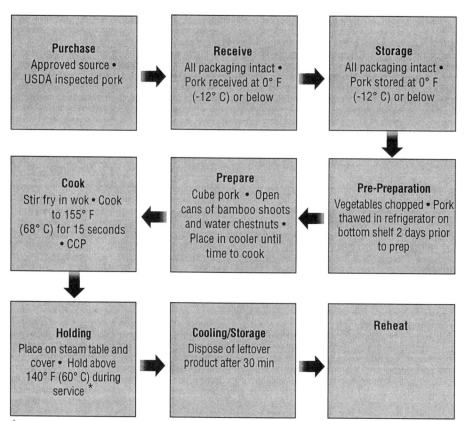

Purchase
Approved source •
USDA inspected pork

Receive
All packaging intact •
Pork received at 0° F
(-12° C) or below

Storage
All packaging intact •
Pork stored at 0° F
(-12° C) or below

Cook
Stir fry in wok • Cook
to 155° F
(68° C) for 15 seconds
• CCP

Prepare
Cube pork • Open
cans of bamboo shoots
and water chestnuts •
Place in cooler until
time to cook

Pre-Preparation
Vegetables chopped • Pork
thawed in refrigerator on
bottom shelf 2 days prior
to prep

Holding
Place on steam table and
cover • Hold above
140° F (60° C) during
service *

Cooling/Storage
Dispose of leftover
product after 30 min

Reheat

* not a CCP since only being held 30 minutes

Chapter 5

Robin should be sent home. Robin's acute symptoms present a risk to food safety. If Robin is allowed to work, she could transmit her illness to a customer and cause a foodborne illness. Janet could continue working, provided that she has clean bandage over the laceration and wears a glove over the bandage.

Chapter 6

The pan used to hold the raw chicken should not have been re-used for the cooked chicken. The cooked chicken could become contaminated by the pan. The second employee failed to properly sanitize the blender. In both cases, the risk was cross contamination because of using unsanitized equipment. The first employee should have obtained clean pans for the cooked chicken. The second employee should have correctly cleaned and sanitized the blender, realizing the difference between clean and sanitary.

Chapter 7

A preventive maintenance program could help to prevent the refrigerator breakdown. Henry needs to discuss equipment maintenance responsibilities with the manager of Engineering and Maintenance. The improperly working oven could result in food not being cooked properly and cause foodborne illness. Henry can justify the purchase of new cutting boards, because the nicked cutting boards could create conditions for bacterial growth, and the financial cost of foodborne illness is much greater than that of replacing worn equipment.

Chapter 8

Contributing causes of the accident include: Juan's shoes, being in a hurry, and the water on the floor. Non-skid shoes, being more careful, keeping the floor mopped and dry, and using a "wet floor" sign might have prevented the accident. Dishroom staff should be trained to clean up spills and use "wet floor" signs, Juan should purchase non-skid shoes, and Juan should be coached on exercising caution during busy times.

Chapter 9

Lynn could have conducted self-inspections to identify potential problems. Lynn could have reviewed safe food handling practices — such as food storage — with staff through training programs. Lynn should immediately correct the problems identified by the inspector, before he/she leaves the operation. Lynn should review the violations with staff and develop a plan to avoid future violations.

Appendix C
Review Questions Answer Key

Chapter 1		Chapter 2		Chapter 3		Chapter 4		Chapter 5	
1.	A	1.	B	1.	B	1.	C	1.	A
2.	D	2.	A	2.	C	2.	A	2.	A
3.	C	3.	D	3.	D	3.	C	3.	D
4.	C	4.	C	4.	D	4.	C	4.	D
5.	D	5.	B	5.	B	5.	B	5.	B
6.	C	6.	C	6.	A	6.	D	6.	B
7.	B	7.	C	7.	C	7.	A	7.	D
8.	C	8.	C	8.	B	8.	D	8.	C
9.	B	9.	C	9.	B	9.	B	9.	B
10.	D	10.	C	10.	C	10.	D	10.	D

Chapter 6		Chapter 7		Chapter 8		Chapter 9	
1.	C	1.	B	1.	B	1.	C
2.	C	2.	B	2.	D	2.	A
3.	C	3.	B	3.	B	3.	B
4.	A	4.	B	4.	C	4.	D
5.	A	5.	C	5.	A	5.	C
6.	A	6.	B	6.	C	6.	C
7.	C	7.	D	7.	B	7.	B
8.	B	8.	B	8.	A	8.	A
9.	C	9.	D	9.	B	9.	B
10.	C	10.	B	10.	A	10.	D

Appendix D
References

Allen, S.D. 1996. *CDM Exam Study Guide*. Itasca, IL: Dietary Managers Association.

Allen, S.D. 1997. *Food Protection Training Program*. Itasca, IL: Dietary Managers Association.

Avery, A. 1985. *A Modern Guide to Food Service Equipment*. New York: Van Nostrand Reinhold Company.

Baltzer, L.E., PhD, RD, LD, and Gilmore, S.A., PhD, RD, LD 1992. *Food Preparation Study Course — Quantity Preparation and Scientific Principles*. Third Edition. Ames, Iowa: Iowa State University Press.

Birchfield, J. C., 1988. *Design and Layout of Foodservice Facilities*. New York: Van Nostrand Reinhold Company.

Bostic, J.L., MPH, RD, and Lavella, B.W., MBA, RD, 1997. HACCP *for Food Service Professionals*. 2nd Edition. St. Louis, MO: Lavella Food Specialists.

Bryan, F.L., PhD, MPH (July/August 1988). Hazard Analysis Critical Control Point: What the system is and what it is not. *Journal of Environmental Health*, 50:7, 400-401.

Cichy, R.F. 1993. *Sanitation Management*. East Lansing, Michigan: Educational Foundation of the American Hotel and Motel Association.

Cody, M. M., PhD, RD and Kieth, M, PhD, RD. 1991. *Food Safety for Professionals: A Reference and Study Guide*. Chicago, IL: American Dietetic Association.

Cohen, G. and Cohen N. 1982. *Food Service Sanitation Handbook*. Rochelle Park, New Jersey: Hayden Books.

Cooperative Extension Service of the University of Georgia. 1992. *Cleaning, Sanitizing and Pest Control in Food Processing, Storage, and Service Areas — Bulletin 927*. Athens, Georgia: Georgia Extension Service.

Cooperative Extension Service of the University of Georgia. 1993. *Maintaining Food Quality in Storage — Bulletin 914*. Athens, Georgia: Georgia Extension Service.

The Educational Foundation of the National Restaurant Association. 1993. The *HACCP Reference Manual*. Chicago, IL: National Restaurant Association.

Gisslen, W. 1989. 2nd Edition. *Professional Cooking*. New York: John Wiley and Sons.

Guthrie, R. K. 1988. *Food Sanitation*. Third Edition. New York: Van Nostrand Reinhold Company.

Kinneer, J. 1997. *Food Protection Study Guide*. Itasca, IL: Dietary Managers Association.

Kinneer, J. 1996. *Food Safety Skillbook*. Itasca, IL: Dietary Managers Association.

Kinneer, J. (October 1997). Job instructional training and in-service training — important elements of an effective staff education program. *Dietary Manager Magazine*, p.26.

Kotschevar, L.H., and Donnelly, R. 1993. *Quantity Food Purchasing*. Fourth Edition. New York: Macmillan Publishing Company.

LaVella, B., MBA, RD. (November/December 1995). How to use HACCP in your facility. *Dietary Manager Magazine*, 10-12.

Loken, J.K., CFE. 1994. *HACCP Food Safety Manual.* New York: John Wiley and Sons.

Marriot, N.G., (1989). *Principles of Food Sanitation.* 2nd edition. New York: Van Nostrand Reinhold Company.

Minor, L.J. and Cichy R., 1984. *Foodservice Systems Management.* West Port, Connecticut: The AVI Publishing Company, Inc.

Morgan, W.J. , 1974. *Supervision and Management of Quantity Food Preparation.* Berkley, CA: McCrutchan Publishing Corporation.

National Restaurant Association. 1993. *Serving Safe Food.* Chicago, IL.

Pierson, M.D., and Corlett, D.A., Jr. 1992. *HACCP Principles and Applications.* New York: AVI Book, Van Nostrand Reinhold Company.

Puckett, R.P., MA, RD, LD and Norton, Charnette L., MS, RD, LD, FADA. 1996. 2nd edition. *HACCP — The future challenge.* Missouri City, TX: The Norton Group, Inc.

Puckett, R., MA, RD, LD and Ninemyer, J., Ph.D., 1992. *Managing Food Service Operations.* Second edition. Lombard, IL: Dietary Managers Association.

Sambataro, P. (November/December, 1995). HACCP Implementation — An effective way to ensure food safety. *Dietary Manager Magazine*, 6-9.

Spears, M.C., 1995. 3rd edition. *Food Service Organizations, A Managerial and Systems Approach.* Englewood Cliffs, NJ: Prentice-Hall, Inc.

Stevenson, K., PhD and Bernard, D. T. (editors). 1995. *HACCP — A workshop manual.* Washington, DC: The Food Processors Institute.

United States Public Health Service. *Food Code 1997.* Washington, DC: Government Printing Office.

Index

A

accidents. *See also* safety of foodservice
    employees
accidents, foodservice
  investigating 143
  investigation form, *figure* 147
  types and prevention 141
acidity 6
air gap 122
Anisakis 13
aprons, changing 90
available water 6

B

Bacillus cereus 9
back-siphonage 122
backflow 122
bacteria
  aerobic 5
  anaerobic 5
  conditions for growth 6
  facultative 5
  spores 5
bacteria. *See also* pathogens
burns. *See* accidents, foodservice: types and
    prevention

C

Campylobacter jejuni 9
CAP. *See* controlled atmosphere packaged foods
carrier 7
CCP. *See* critical control point
CDC. *See* Centers for Disease Control and
    Prevention
ceilings. *See* construction materials
Centers for Disease Control and Prevention 161
chemical test strips
  for sanitizing solutions 104
chemicals, safe storage 108
ciguatera 15
cleaning
  definition 101
  manual 104
    steps, *figure* 109
  mechanical 106
    temperatures, *figure* 110
cleaning products 100
  as chemical hazards 15
  types of detergents 106
cleaning program 100
  sample schedule 102
  self-inspection checklist, *figure* 103
Clostridium botulinum 9

Clostridium perfringens 10
clothing
for foodservice workers 90
construction materials 120
controlled atmosphere packaged foods 30
receiving 32
cooking
microwave 47
temperature log, *figure* 46
thoroughly to prevent illness 44
time and temperature guide 45
cooling foods 51
as a critical control point 62
methods, *figure* 52
crisis 166
communication during 169
examples 166
crisis management plan 167
critical control points 62. *See* Hazard Analysis Critical Control
Point (HACCP): critical control points
cross contamination 16, 43
during storage 33
during temperature measurement 129
prevention 43
cross-connection 122
cuts. *See* accidents, foodservice: types and prevention
cutting boards 126

D

danger zone. *See* hazard zone
detergents. *See* cleaning products
diarrhea
as symptom of foodborne illness 9-14
restrictions, *figure* 92
disease transmission
by employees 93
figure 91
restrictions for preventing, *figure* 92
dishes, storage 108
dishmachine. *See* cleaning, mechanical
dishwashing. *See* cleaning, manual; cleaning, mechanical
disposable gloves 89
dry lab 67

E

electrical shocks. *See* accidents, foodservice: types and
prevention
equipment. *See also* purchasing equipment
clean-in-place 130
cleanability 120
installation 126
maintenance 132
checklist, *figure* 133
specifications 132
Escherichia coli 0157:H7 10

F

falls. *See* accidents, foodservice: types and prevention
FDA. *See* Food and Drug Administration
FDA Model Food Code. *See* Model Food Code
FIFO. *See* first in, first out
fire extinguishers 146
fire safety. *See* safety, fire
first in, first out 33
floor drains 121
floors. *See* construction materials
flow of food 42
Food and Drug Administration 160
food preparation areas, construction 123
food purchasing. *See* purchasing food
food safety
importance of cleanliness 99
manager's responsibilities 81
foodborne illness
causes 2
definition 1
frequency 2
outbreak 1
management of 168
prevention 16
foodborne infection 5
foodborne intoxication 5
foodservice practices
for safe food 90
fungi 8

G

grading of food 26

H

HACCP. *See* Hazard Analysis Critical Control Point
HACCP-based inspection. *See* inspections of foodservice
    operations: HACCP-based
hair restraints 90
hand washing
    determining compliance 87
    facilities, construction 123
    sinks and supplies 86
    technique 89
        figure 88
    use of nail brush, *figure* 87
    when to do 87
Hazard Analysis Critical Control Point (HACCP) 59
    advantages 60
    analyzing hazards 60
    control points, monitoring 65
    corrective actions 66
    critical control points 62
    critical control points, *figure* 64
    critical limits 65
        examples, *figure* 65
    flow chart, *figure* 68
    identifying risks 62
    implementing 60
    logs 67
    plan, example 67
    plan, example, *figure* 69
    recipe, example 72
    record-keeping system 67
    seven principles, *figure* 63
    steps 60
    terms, *figure* 61
Hazard Communication Standard 150
hazard zone
    definition 6
    figure 48

hazards
    biological 2, 3
    chemical 2, 15
    physical 3, 16
heating, ventilation, and air conditioning 121
Hepatitis A 14, 93
    restrictions, *figure* 92
HIV/AIDS 93
    restrictions, *figure* 92
holding food
    cold 47
    handling utensils 51
    hot 47
    temperature log for, *figure* 50
host 8
HVAC. *See* heating, ventilation, and air conditioning
hygiene 85
hygiene. *See also* disposable gloves; hair restraints; hand
        washing
hygiene practices 85

I

ice, food contact with 48
illness, restrictions, *figure* 92
infrared sensors for hand sinks 86
injury. *See* accidents, foodservice
inspection of food 25
inspections of foodservice operations 163
    areas addressed 163
    dealing with violations 164
    HACCP-based 165
    preparation 165
integrated pest management. *See* pest control

J

jewelry 89

L

laws and regulations
    affecting foodservice operations 159
    differing 159
    Federal 160
    local 159
    state 159
licensed pest control operator 112
lighting for foodservice facilities 122
Listeria monocytogenes 11

M

MAP. *See* modified atmosphere packaged foods
Material Safety Data Sheet 152
    figure 151
metals 15
microwave cooking. *See* cooking, microwave
Model Food Code 162
modified atmosphere packaged foods 30
    receiving 32
molds 8
    guidelines for handling 8
MSDS. *See* Material Safety Data Sheet
mycotoxins 14

N

National Marine Fisheries Service 161
NMFS. *See* National Marine Fisheries Service
Norwalk virus 14, 93
NSF seal 124

O

occupational illness 143
Occupational Safety and Health Administration 149
    inspections 149
Occupational Safety and Health Act 149
OSHA. *See* Occupational Safety and Health Administration

P

parasites 8

parasites. *See also* Anisakis, Trichinella spiralis
pathogens 4
    common foodborne, *figure* 9
personal hygiene. *See* hygiene, hand washing
pest control 109, 111
    integrated pest management 112
pesticides 15, 111
pests in foodservice operations 109
    birds 111
    cockroaches 110
    houseflies 111
    rats and mice 111
    pH range 6
physical facilities
    impact on food safety 119
plumbing 122
plumbing. *See also* back-siphonage; backflow; cross-connection
plumbing system 122
potentially hazardous foods
    as a condition for bacterial growth 6
    definition 5
    in HACCP 60, 66
    in hazard zone 7
    receiving 28
POTTWA, conditions for bacterial growth 6
prepared foods, storage 53
Public Health Service 25
purchasing equipment 124
    considerations 120, 125
    refrigeration 125
purchasing food
    from approved sources 25
    guidelines 25
    vendor selection 26

R

receiving food 28
    as a control point 67
    checklist 29
    quality indicators, *figure* 31
    verifying food temperatures, *figure* 30

refrigeration equipment. *See* purchasing equipment: refrigeration
regulations. *See* laws and regulations
reheating food 53
respiratory infection restrictions, *figure* 92

S

safety, fire 146
    self-inspection checklist, *figure* 148
safety of foodservice employees 141
    management principles 141
    safe work practices, *figure* 144
    self-inspection checklist, *figure* 142
Salmonella enteritidis 11
sanitary
    definition 101
sanitizing
    with chemicals 104
        safe storage of products 108
        usage guidelines, *figure* 107
    with heat 104
scromboid poisoning 15
self-service food bars 51
shellfish identification tags 25
Shigella 12
skin lesions
    precautions 89
    restrictions, *figure* 92
slips. *See* accidents, foodservice: types and prevention
sprains. *See* accidents, foodservice: types and prevention
Staphylococcus aureus 12, 93
storage of food 32
    facilities 123
    guidelines, *figure* 35
    prepared 53
    temperatures 33, 36
strains. *See* accidents, foodservice: types and prevention
Strep throat restrictions, *figure* 92

T

tasting food 90

temperature
    log for holding food, *figure* 50
    measuring in cooked food 45
    monitoring 43
    monitoring, *figure* 49, 129, 130
thawing methods 44
thermocouple probes 127
thermometers 127
    calibrating, *figure* 128
time
    as a control technique 43
time/temperature indicator 32, 127
toxins 15
trainer selection 82
training 81
    essential to food safety 80
    group 82
        examples, *figure* 84
    on-the-job 82
        procedure, *figure* 83
    orientation 81
    resources 84
transport of food 51
Trichinella spiralis 13
trips. *See* accidents, foodservice: types and prevention

U

U.S. Department of Agriculture 160
    grading of meat 26
    standards for meat, poultry, and eggs 25
UHT 32
UL mark 124
ultra-pasteurization 32
utensils, using 90, 91
Underwriters' Laboratories. *See* UL mark
USDA. *See* U.S. Department of Agriculture

V

viruses 7, 93
viruses. *See also* Hepatitis A, Norwalk virus

W

walls. *See* construction materials
warewashing. *See* cleaning, manual
  equipment 131
warewashing areas, construction 123
waste management 112
water, safe source 18
wiping cloths 101

Y

yeasts 8
Yersinia enterocolitica 13